A Gardener's Encyclopedia of Wildflowers

An Organic Guide to Choosing and Growing Over 150 Beautiful Wildflowers

C. Colston Burrell

Foreword by Ann Lovejoy

Rodale Press, Inc.
Emmaus, Pennsylvania

A FRIEDMAN GROUP BOOK

Published in 1997 by Rodale Press, Inc.

The information in this book has been carefully researched, and all efforts have been made to ensure accuracy. Rodale Press, Inc., assumes no responsibility for any injuries suffered or for damages or losses incurred during the use of or as a result of following this information. It is important to study all directions carefully before taking any action based on the information and advice presented in this book. When using any commercial product, *always* read and follow label directions. Where trade names are used, no discrimination is intended and no endorsement by Rodale Press, Inc., is implied.

A GARDENER'S ENCYCLOPEDIA OF WILDFLOWERS
was prepared and produced by
Michael Friedman Publishing Group, Inc.
15 West 26th Street
New York, New York 10010

Michael Friedman Publishing Group Editorial and Design Staff:

Editor: Susan Lauzau
Art Director: Lynne Yeamans
Designers: Susan E. Livingston and Garrett Schuh
Photography Editors: Samantha Larrance and Karen Barr

Rodale Press Home and Garden Books Staff:

Managing Editor: Ellen Phillips
Editors: Fern Marshall Bradley, Deborah L. Martin
Senior Research Associate: Heidi A. Stonehill
Copy Editor: Erana Bumbardatore
Vice President and Editorial Director: Margaret J. Lydic
Copy Director: Dolores Plikaitis
Associate Art Director: Mary Ellen Fanelli
Office Manager: Karen Earl-Braymer

If you have any questions or comments concerning the editorial content of this book, please write to:

Rodale Press, Inc.
Book Readers' Service
33 East Minor Street
Emmaus, PA 18098

Library of Congress Cataloging-in-Publication Data

Burrell, C. Colston.
A gardener's encyclopedia of wildflowers : an organic guide to choosing and growing over 150 beautiful wildflowers / C. Colston Burrell
p. cm.
"A Friedman Group book"—T.p. verso.
Includes bibliographical references (p.) and index.
ISBN 0–87596–723–X (hardcover : alk. paper)
1. Wild flowers—Encyclopedias. 2. Organic gardening. I. Title.
SB439.B87 1997 96–25130
635.9'5173—dc20

Distributed in the book trade by St. Martin's Press

Color separations by Fine Arts Repro House Co., Ltd.

Printed in the United Kingdom

2 4 6 8 10 9 7 5 3 1 hardcover

OUR PURPOSE

"We inspire and enable people to improve their lives and the world around them."

To my mother, at whose knee I first discovered the beauty of the bloodroot and who taught me to see wildflowers through eyes filled with wonder. And to my father, who taught me to see them through the lens of a camera.

Acknowledgments

I have had many gifted teachers in the course of my life. Mother Nature was, and still remains, my mentor. She teaches me daily lessons subtle and profound. I owe an enormous debt to the early writers who related their knowledge and experience in spellbinding books that have been my companions since the day I discovered them at the local library. In particular, Mrs. William Starr Dana gave flight to my imagination, and Bebe Miles was the wind in my sails. My time spent in the gardens of Mary Corley, Hazel Evans, and Sue Sneed taught me the wisdom that comes from a lifetime of trial, error, and artistic pursuit. Lovinia Callis, Lynda Fleet, and my dear sister Susan were my constant companions on field trips and plant rescues as the suburbs engulfed the forests around homes.

I have many personal friends who have offered advice, toured gardens with me, answered my questions, given lectures, written books, and picked up the other end of the phone when they were needed. I would like to thank Allen Bush, Neil Diboll, Edith Eddleman, Judy Glattstein, Christopher Lloyd, and Barry Yinger for being my teachers and critics. Thanks also go to Janie Bryan, Lynn Cohen, Dale Hendricks, and Bob McCartney for answering questions relating to the text.

Special thanks go to Edith Eddleman and Fred Rozumalski for their outstanding designs, to Ellen Phillips, and to Bruce Ellsworth, who made dinner all those nights while I was at the computer.

Contents

Foreword

Congratulations! You are about to take an in-depth tour of a little-known world, one populated by our native plants. The best and most beautiful candidates from native habitats all across the country have been gathered here for your pleasure. They come from meadow and woodland, prairie and mountain, but all of them are adaptable enough to grow happily in your own yard, supplying your beds and borders with multiple seasons of colorful flowers and foliage.

What's more, you can jump right in and start growing these beautiful wildlings without worries. We often hear warnings about not digging plants from the wild because most of them will die in captivity. However, when we plant nursery-propagated wildflowers sold by reputable nurseries (ones you will find recommended in these pages), we are actually helping to encourage and protect these natural resources.

Gardeners who delight in ornamental border beauties will discover a wide range of native dazzlers here. Some wild things, like New York aster, went to European finishing school and came home the belle of the ball. Bright creatures like this long-bloomer will decorate island beds or mingle in mixed borders with ease.

Wildflowers such as white trillium are more like the girl next door—her face grows lovelier the better you know her. Many of these subtler wild beauties need thoughtful placement to display their charms to the fullest. Cole Burrell provides you with plenty of guidance in how and where to place native plants to optimal advantage. Indeed, this handbook is packed with terrific ways to use this new cast of garden characters.

Folks who are most interested in creating a healthy, wholesome environment for native plants will learn how to put together plantings that also supply food and shelter to birds and wildlife. Others may want to use native plants as groundcovers in shady woodland gardens or around shrubs.

Perhaps you are looking for adaptable, hardy border plants that don't need much fussing over, or you want sturdy perennials that are less finicky than those expensive English imports. If so, you have come to the right place.There could be no better guide to wildflowers than Cole Burrell, whose deep affection for our native flora has led him to develop garden designs that are both beautiful and ecologically sound. An extraordinary plantsman, Cole has the experience and skill to combine natives with border exotics effectively, creating visually satisfying combinations while adapting planting patterns that echo natural relationships found in woods and meadows across the country.

His chapters on gardening with wildflowers are unique in offering designs for entire yards as well as individual plantings. These practical and adaptable plans will be extremely helpful for gardeners who want to achieve a consistent look throughout the whole garden or property. For those who want to take things slowly, adding sections of wildflowers as confidence and skills build, suggestions are offered for scaling down the larger projects or implementing them one area at a time.

Whether you decide to transform the whole yard or begin with a single wildflower bed, you will find all the information and inspiration you need in the following pages. Enjoy the journey!

Ann Lovejoy
Bainbridge Island, Washington

Preface

Though my life has been filled with teachers, I am by and large self-taught. I am self-taught in that I learned about wildflowers on the banks of clear streams and on the slopes of verdant mountains. I helped my mother plant wildflowers in her first garden as soon as I could hold a trowel. We learned together, as she was a city girl who followed the American dream to the suburbs in the 1950s. I had my own little patch of wildflowers by the time I was 9, and by 14, our acre of land was too small for all the plants I wanted to grow.

When it came time for college, I knew I wanted to study plants. That simple goal has led me through four degrees, but it was the rare class—and rarer professor—that taught me a fraction of what I had learned outdoors. Not that I didn't learn plenty in school. After 14 years of college, you would expect a few things to sink in. I learned how plants got their names, and why people keep changing them. I learned how and why the things I had observed were so. I learned about food chains, environmental interrelationships, design theory, site engineering, and soil chemistry, but every moment I wasn't in the classroom, I was in the garden or in the wild. I tended my first garden for over 20 years, and it taught me lessons that no curriculum or expert possibly could.

I have a quarter-acre plot of earth now. It lies practically in the shadow of the great skyscrapers of Minneapolis. My garden is only five years old, yet it seems timeless. It is my retreat, my laboratory, my photography studio, and my playground. I have tried to accommodate as much wildlife as I can by designing the garden like a native plant community. Not all of the 1500 species and cultivars that grow there are native, but at least half of them are. My garden meets not only my needs, but the needs of birds, butterflies, and other wildlife as well. I have seen over 120 different bird species in the garden. At last count, nine different butterflies have stopped to visit my flowers since I planted them. Before I started gardening, I saw no butterflies here. Through proper design and planting, we can give a home to creatures that are often shut out of our cities and suburbs. We are all the richer for it.

Part I

Welcome to Wildflower Gardening

◀ *Colorful wildflowers are at home in the garden and in their natural settings. These prairie beauties—goldenrod, purple-spired gayfeathers, and the striking greenish-white flowerheads of rattlesnake master—would look just as good in a sunny garden.*

Why Grow Wildflowers?

Wildflowers are one of the loveliest parts of the natural landscape. Many are also well suited to our gardens. You can find wildflowers in every part of the United States, gracing fields, woodlands, and wetlands, along beaches, and on rocky mountain slopes. Sweeps of bluebells and trout lilies carpet the ground on a woodland floor, Joe-Pye weed and asters enliven a moist roadside. In the spring, even the deserts of the Southwest come alive with flowers, while in the fall, eastern meadows are awash with colorful goldenrods and asters.

It's easy to grow many of these lovely wildflowers in your flowerbeds and borders. When matched to the soil and light conditions of their wild homes, wildflowers bloom prolifically and suffer few insect and disease problems. If you've tried growing a few wildflowers, you're probably enchanted with their beauty and easy care. You may be ready to try creating a whole garden that focuses on wildflowers, whether it's a jewel-like patch of meadow wildflowers or a woodland garden under your shade trees.

Native Wildflowers

Every area has a group of plants and animals that have lived there naturally for thousands of years. These are that area's native flora and fauna. A plant can be native to a region, state, or just a certain valley. Native plants are what give an area its regional identity—they help make it look different from other areas.

Butterfly weed, Joe-Pye weed, asters, and garden phlox are just a few of the native wildflowers that have also become well-loved garden flowers. These flowers are native to particular areas in the United States and Canada, but they've become so popular that you'll see them in gardens all over North America—including many regions where they are not originally native.

This book is an encyclopedia of wildflowers that are native to the United States. The techniques for growing and propagating wildflowers described here will work equally well for both native and introduced plants.

Introduced Wildflowers

Introduced wildflowers are plants that grow wild in areas where they are not native. Many beautiful and popular "wildflowers," like the stunning red Shirley poppies, are actually plants introduced to North America from other continents. Most are European natives brought here by settlers for food, medicine, or ornament. Other plants came by chance; their seeds may have been mixed in with agricultural seeds or even been part of the bedding used on ships that crossed the Atlantic on their way to the new colonies. These imports include cheerful roadside flowers such as oxeye daisy and yarrow. They are now so familiar that we think of them as American wildflowers.

Over 20 percent of the plants listed in Peterson's *Field Guide to Wildflowers* are introduced plants, brought to North America from Europe and Asia. Introduced plants are

Butterfly weed (Asclepias tuberosa) is a showy roadside wildflower native throughout eastern and central North America.

Oxeye daisy (Chrysanthemum leucanthemum) is a European plant that now grows wild in most parts of North America.

Lovely Queen-Anne's-lace (Daucus carota) is a common sight on roadsides and in meadow gardens, but it is native to Europe, not North America.

Pass up the temptation to grow purple loosestrife (Lythrum salicaria). It's highly invasive and displaces native wildflowers.

sometimes called exotics or aliens. Some of them—such as chicory and Queen-Anne's-lace—are often pictured in magazine articles about native wildflowers, creating confusion about the true meaning of the term "native."

Invasive Plants

Which will grow better on a given site—a plant that's native to the site or an introduced plant? It might seem logical that natives would grow best because they have adapted to the site. This isn't always true, though. I can think of dozens of nonnative species, such as oriental bittersweet, that thrive in the United States although they did not evolve here. These invasive plants succeed, and even overcome the native plants, because they evolved under similar conditions on another continent and have left their natural insect predators behind them. Few plants are better adapted to our landscape than Japanese honeysuckle and kudzu, although neither one is native to North America.

While many introduced plants are harmless and beautiful additions to gardens and roadsides of their adopted shores, others collide head-on with the local flora. For example, many wetlands in northern North America are now overrun by purple loosestrife, a perennial plant native to Europe. This plant was propagated in nurseries and grown in gardens for years before naturalists and horticulturists realized how aggressively it invaded natural areas, crowding out the native plants.

Exotic thugs such as buckthorn, porcelain berry, Japanese knotweed, and Norway maple are outcompeting native species and reducing the number of plant species that grow naturally throughout the eastern and central states.

Types of Wildflower Gardens

Thanks to a greater public appreciation of wildflowers and improved propagation techniques, a dazzling array of native wildflowers is now available to home gardeners. With this tide of new plants have come new ways of gardening with wildflowers. Gardeners combine them with plants from all over the world to create distinctive gardens. Rather than creating separate wildflower areas, gardeners have planted native wildflowers in their traditional flowerbeds and borders. The focus has shifted from exclusively woodland gardens—and now includes informal shade gardens and formal borders. Meadow and prairie gardening for beauty and low maintenance is a popular trend. And gardeners now know that they don't need a field or woods to plant a wildflower garden—you can have a postage-stamp wildflower meadow or a beautiful woodland garden under a single tree.

The three major types of wildflower gardens discussed in this book are woodland, meadow, and prairie gardens. In a woodland garden, you'll discover a constantly changing garden picture composed of flowering plants, grasses, shrubs, and trees. In a meadow, wildflowers bloom in a seasonal progression that ends with an autumn blaze of golden yellows, deep blues, and rich purples. Prairie gardens mix flower colors with the greens and golds of grasses but seem wonderfully alive as a mix of fine-textured grasses sways in the breeze. The textures, colors, and forms of evergreen and dried wildflower foliage, dried grasses, and ripened seedheads keep wildflower gardens beautiful even in winter.

Try combining native wildflowers like asters and goldenrod with more traditional perennials like sedum to create a stunning meadowlike garden.

The red flowers of bee balm (Monarda didyma) are standouts in this garden of native wildflowers.

When you design your own wildflower garden, your goal should be to capture the feeling of a natural woodland, meadow, or prairie. You can accomplish this in one of two ways. The first is to use only the plants native to North America, to your state, or even to your local area.

The second approach is to use native wildflowers along with plants from many other countries but still choose plants that are well adapted to the kind of garden you're planting. For example, you wouldn't plant a prairie grass in your woodland garden, whether it was native or not. If you're using nonnative plants in a wildflower garden, be sure to choose species and cultivars that are hardy in your area. Make sure you don't plant aggressive wildflowers that may overrun your garden and surrounding areas.

You can add extra interest and enjoyment to a wildflower garden by adding water. Small streams sometimes flow through natural prairies and meadows, and wet meadows include boggy areas. If you're interested in woodland gardens, considering adding a small water garden to make your garden extra special. You'll find several wildflowers in the encyclopedia that thrive in boggy conditions or along streambanks. For how-to references on constructing a water garden, refer to "Recommended Reading" on page 184.

WOODLAND GARDENS

Each year, the deciduous forests of eastern and central North America put on a gorgeous display of wildflowers. From late March through May, almost all the wildflowers bloom in the spring sunshine. Woodland wildflowers produce such a dramatic show because their bloom cycles are tied to the availability of light and moisture. In the spring, woodlands are moist from winter rain and snow, and because the trees are bare, the woodland floor is flooded with light. By the time summer arrives, the canopy of tree leaves robs the forest floor of all but filtered light. To survive, woodland wildflowers grow and bloom early in the spring. For gardeners, this means a glorious flower display, but one that may be short-lived. Many of our early spring wildflowers are ephemeral (a word that originally meant "lasting a day") and go dormant soon after blooming.

A groundcover of Allegheny foamflower (Tiarella cordifolia) can brighten a dark area under large trees.

Crested iris (Iris cristata) thrives in a woodland setting.

Other spring wildflowers, like foamflowers, keep their foliage all through the summer and even turn bright colors in the autumn. They add to the foliage and fruit displays of the woody plants above them. Late-blooming woodland plants such as asters and goldenrods produce a flush of lush foliage early in the year. They spread broad leaves and grow steadily through the year. As summer wanes, they bloom in white, gold, and purple, giving a signal of the autumnal spectacle to come.

A woodland garden under evergreens like pines or hemlocks will look quite different than a woodland garden under deciduous trees. Wildflowers native to coniferous forests have adapted to low light levels year-round. Also, evergreen needles are effective rain barriers. Much of the precipitation in northern and western forests evaporates without ever reaching the forest floor. To survive in continuous shade, plants develop thick, evergreen leaves. Plants such as round-lobed hepatica, partridgeberry, and bunchberry have evergreen leaves that can photosynthesize in early spring and fall when other plants are dormant. Evergreen leaves are tough and leathery so they can endure dry spells.

When you design a woodland garden, be sure to include layers of plants. In a natural woodland, wildflowers are at ground level, often raising their foliage only a few inches above the leaf litter. As spring ephemerals go dor-

Native phlox and columbine mix beautifully with other perennials and bulbs in this woodland garden.

mant, ferns, great merrybells, and other plants with long-lasting foliage take their place. Above them, shrubs grow in patches where light and space are sufficient. The next layer, the understory, consists of small trees, while the tallest layer, the canopy, covers the entire garden. Don't be intimidated by all this layering—you can create a lovely woodland garden under a single large shade tree by planting smaller trees, shrubs, and wildflowers beneath it.

MEADOW GARDENS

Meadow gardens are popular for a good reason. Wildflowers fill meadows with color and fragrance. Grasses have wonderful textures to look at and touch, and their soft swishing sounds are a soothing change from the noise and bustle of cars, computers, and televisions.

Native meadow wildflowers produce flowers mostly in summer and fall. There are two types of native grasses:

This moist summer meadow garden boasts a riotous display of ironweed (Vernonia noveboracensis) and goldenrods.

This sunny meadow is alight with the yellow daisies of black-eyed Susan (Rudbeckia hirta) and, behind them, the taller Mexican hat (Ratibida columnifera). Purple coneflower (Echinacea purpurea) at the front of the picture and wild bergamot (Monarda fistulosa) in the center add lavender-pink accents.

cool-season and warm-season. Cool-season grasses like tufted hairgrass begin growth early and flower in spring or summer, then they go dormant during the heat of midsummer. Warm-season grasses like big bluestem and Indian grass begin growth later in the season, after temperatures moderate and days lengthen. The warm-season grasses go dormant in the fall like garden perennials.

Natural meadows grow in breaks in the forest where sunlight reaches the ground all through the growing season. Most meadows contain both grasses and wildflowers. In the midwestern and eastern states, the balance is tipped in favor of grasses.

Another kind of meadow develops in abandoned farm fields. When a farm field is abandoned, "pioneer" species move in—a mix of native and naturalized annuals such as ragweed, goldenrod, and Queen-Anne's-lace that colonize open or disturbed ground. Over time, an increasing variety of plants such as aster, milkweeds, and grasses parade across the landscape. The area will eventually return to forest unless there is a fire or human intervention. In your meadow garden, you'll need to stop this natural process of change—especially the sprouting of woody plants—by periodically mowing or burning the meadow.

When you plant a meadow, you'll need to decide how closely you want to imitate nature. Will you create a garden of meadow flowers or re-create a meadow habitat? The best way to decide what kind of meadow garden you want is to study the natural meadows in your area. You'll

Texas bluebonnets are a familiar sight on prairies in the Texas hill country. These cheerful annuals will self-sow in your prairie garden.

find, for example, that black-eyed Susans and butterfly weed grow and bloom in old-field meadows. New England meadows are full of lilies, ironweed, and asters.

It's important to buy plants for your meadow garden from local nurseries that produce their own stock because meadow plants of the same species can differ from area to area. For example, switchgrass from Nebraska is taller and coarser than switchgrass from Minnesota. When planted in Minnesota, the Nebraska plants look out of place and may cross-pollinate with local plants, thereby changing the appearance and the genetic makeup of the Minnesota plants. This could result in changes in flowering time, hardiness, susceptibility to diseases, or aggressiveness. The same is true for prairie gardens.

PRAIRIE GARDENS

A natural prairie is an open grassland, not a woodland opening like a meadow although they may contain many of the same plants. Prairies are nearly treeless and may have scattered patches of shrubs. In the spring, prairie grasses are low and cushionlike. As the weather warms, the spiky leaves and flowerstalks push upward. In summer and early autumn, flower plumes and spikes spread puffs of cottony seed across the prairie. Wildflowers bloom and add color throughout the season, ending with a blaze of gold and purple. Even in winter, prairies have an enchanting beauty. They also provide cover for wildlife, and songbirds feast on the abundant seeds.

Grasses thrive in prairies because they have extensive underground networks of fibrous roots that probe deeply for moisture. These plants produce foliage and flowers in response to available moisture. In times of plenty, growth and flowering are lush. During drought, plants may remain semidormant to conserve resources.

Where prairies meet forests, there are open, parklike transition areas called savannas. Savannas are a mixture of small expanses of prairie and open groves of trees. Grasses and sun-loving wildflowers grow intermingled with woodland plants. It's easy to convert a treeless area of your lawn into a prairie or to turn an area between trees into a savanna garden.

Historically, fires burned vast tracts of prairie land. Woody trees and shrubs couldn't withstand the intense heat of these blazes, but wildflowers and grasses were able to resprout after the fires because their crowns are below the soil surface where they are protected from the heat. In this way, fire was an important force in maintaining the prairie. In a small prairie garden, you can use your lawnmower instead of fire to keep down woody growth so that your prairie grasses and wildflowers can thrive.

Growing Wildflowers

Because wildflower gardens mimic nature more closely than gardens of cultivated perennials and annuals, they

Purple asters glow in the prairie in fall, contrasting with the seedheads of the grasses. Accent your prairie garden with asters, goldenrod, Joe-Pye weed, and other colorful wildflowers.

In early spring, wet, sunny stream banks are lined with the cheery yellow flowers of marsh marigold (Caltha palustris) and the spiky leaves of iris.

tend to be more trouble-free and easy-care than traditional home gardens. However, even wildflower gardens need occasional weeding, watering, and trimming to grow and look their best.

To succeed with wildflower gardening, learn about the native environments of the plants you want to grow. Once you do, you'll understand why meadow plants will languish in a shaded garden, and plants from cool mountain woods will wither in hot, sunny gardens. The essential rule for a good wildflower garden is to match the plants to the site. The right light, moisture, and soil type are essential considerations. Some wildflowers tolerate a wide range of conditions, while others grow only in a limited range. You can check the conditions that individual wildflowers need by referring to the encyclopedia entries of this book.

If you can match light and moisture conditions, you can amend the soil to grow a wide variety of species. Woodland plants like moist, humus-rich soils. They are perfect for growing in a shade garden with hostas and other garden favorites. Plants of coniferous forests need moist soil that is acidic, and they tend to be the least tolerant of other conditions. Don't try to grow them unless you have a shady site with moist, acid soil.

Prairie plants need deep loamy soils, while many meadow wildflowers can survive in thin, poor-to-average soils. You can use the lovely species from full-sun habitats in formal and informal perennial gardens, which also need full sun. Mountain soils are often thin and rocky on exposed slopes and humus-rich under trees. So, if you want to try growing wildflowers native to mountain slopes, you'll need to prepare a special site for a rock garden.

Buying Wildflowers

Wildflowers are available as bareroot or potted plants from nurseries across the country. While most of these nurseries

Commercial nurseries propagate wildflowers by raising seedlings and rooting cuttings, which they may keep in greenhouses until they are large enough to plant outdoors in the garden.

propagate the plants they sell, some offer plants collected from the wild. I urge you to be a careful buyer and not buy wild-collected plants. Wild-collection robs natural areas of their beauty and can bring rare plants to the brink of extinction.

Sometimes it's not easy to tell whether plants have been dug from wild areas. Don't rely on labels alone! Industry standards allow nurseries to hold wild-collected plants for one or two years and then sell them as "nursery-grown." Make sure the plants you buy are nursery-propagated, not nursery-grown.

Try moss phlox (Phlox subulata) for a bright accent atop rock walls or on dry, sunny banks where few other plants will thrive.

How can you tell if a plant is wild-collected? Be wary of inexpensive plants or large quantity discounts offered by mail-order nurseries. The nurseries have most likely collected these plants from wild areas. You should expect to pay the same price for a well-grown wildflower as you would for a garden perennial like a daylily or hosta. The most widely collected wildflowers are woodland plants such as trilliums, native lilies and orchids, and evergreen wild gingers. Propagating these plants takes a lot of time and effort, so nurseries reap larger profits by selling wild-collected plants. Never buy any native orchids such as lady-slippers; they are impossible to propagate in commercial quantities and are sure to be wild-collected.

BUYING WILDFLOWER PLANTS

Buying wildflowers in containers at nurseries and garden centers is convenient, and it is also a good way to get quality plants. However, you'll have a much broader range of choices if you buy from mail-order companies. Ordering plants through the mail is fun. Browsing through catalogs keeps winter-weary gardeners from going mad. Mail order is also the *only* way to get certain plants. My favorite wildflower nurseries are listed in the Resources section on page 185.

Some gardeners are reluctant to order plants by mail. They fear the plants won't survive the rigors of being shipped in boxes from the nursery to their home. This is not so! Just follow a few simple rules when shopping by mail, and you'll enjoy great success.

The first rule is to order early. This will ensure that you get the best selection. Next, specify a shipping date appropriate to your area of the country. For example, if you live in New York State and you're ordering plants from North Carolina, request that they ship in May. The plants will be growing strongly due to their early start in a warm climate, and they will establish themselves beautifully after planting. However, if you're a southern gardener ordering from a northern nursery, your best bet is to wait and request fall shipment. Otherwise, by the time the nursery can ship you growing plants, it may be too hot or dry for them to establish themselves well in your garden.

Another important rule: Unpack plants as soon as they arrive. Water container-grown plants well. If plants are bareroot, check them over, water if necessary, and keep them cool and shaded until you can plant them.

BUYING SEED MIXES

Growing wildflowers from seed takes more time and effort than buying plants, but it's much less expensive. You can buy prairie or meadow seed mixes especially prepared for your location, soil, and moisture conditions.

Read the labels on wildflower seed mixes carefully. Some mixes touted as regionally native actually contain a majority of plants that are not even native to North America, much less the region for which the mix is intended. To be sure you have a native mix, order from a specialty native plant nursery like the ones listed in the Resources section on page 185. If you're adventurous and willing to do some research, you can create your own custom mix. (Some nurseries will create custom mixes for you as well.) Whether you buy seeds at a nursery or order them by mail, refrigerate them until it is time to sow them.

Beware of highly publicized meadow seed mixes sold in cans. The labels on these products promise sweeps of flowers simply by scratching the soil and throwing down the seeds. Claims of no maintenance add to the temptation to try them. You won't get any wildflowers just by throwing seed onto your lawn. Unless you have weed-free soil that's already bare, establishing a wildflower garden requires as much initial maintenance as any other garden. But if you'd be happy with a plot of colorful annuals and imported perennials that give the visual effect of a meadow, then try your luck with a canned product. The effects can be very attractive and enjoyable, as long as you know what you are getting and you prepare the site first. Bear in mind that the colorful annuals aren't likely to grow again after the first season. So if you want the same effect every year, you'll have to prepare the site and sow your meadow all over again, just like a big annual garden. You may decide it's better to plant a meadow entirely of perennials that will rebloom every year.

Getting the Garden Ready

Your wildflower garden will need little care once it's established—much less than a traditional annual or perennial garden. But to get to that point, you'll need to start it off right by preparing the bed. The amount of soil preparation needed depends on the kind of wildflower garden you're planning. Woodland gardens need moist, humus-rich soil. For prairies and meadows, the best soil conditions will vary from wet to dry and poor to rich, depending on the kinds of plants you choose to grow.

TESTING THE SOIL

If you think your soil will need enrichment, start by testing it. A soil test will tell you the texture, structure, and pH of the soil as well as the nutrient content. These factors directly influence plant growth. Soil texture refers to the mix of particles that make up the soil. The basic particles are sand (which is large in comparison to the other types of particles), silt, and clay (the smallest particles).

Soil structure refers to the physical arrangement of the soil. Soils with good structure have lots of microscopic pores, or openings, in them that can hold water and air. Soil pH is a measure of how acidic or alkaline the soil is. It's expressed as a number from 1.0 (highly acidic) to 14.0 (highly alkaline). A large selection of plants like to grow in soil that is close to neutral, that is, soil with a pH of 6.5 to 7.0. Many other species require a more acidic soil for best growth. Refer to the encyclopedia entries in this book to learn about pH requirements.

How to test soil. You can have your soil tested through your local Cooperative Extension Service office. The analysis may be free, or it may cost up to about $10. You can also have your soil analyzed by a private soil-testing laboratory.

You'll need to collect soil samples from several spots in your garden. To collect a sample, clear the soil surface and use a stainless steel trowel to dig a hole 4 or 5 inches deep. Then use the trowel to cut a slice of soil from the side of the hole. Mix several samples in a stainless steel or plastic container. Put some of the mixed soil (1 or 2 cups is plenty) in a bag or container provided by the testing service, and mail it.

Soil test reports. The testing service will send you a report about your soil. It will include information about texture and pH, organic matter content, and levels of nitrogen, phosphorus, and potassium. It will also include recommendations about what to apply to your soil to supply any nutrients that are lacking. It's important to request recommendations for organic soil amendments; otherwise, the report may only suggest various kinds of chemical fertilizers. Once you know what your soil needs, you can begin preparing the site.

CLEARING THE SITE

Because you'll be planting under or near trees, you'll have to deal with tree roots in the soil when preparing a site for a woodland garden. Disturb the soil under trees as little as possible because digging can damage active surface tree

To avoid damaging the roots of mature trees, plant woodland wildflowers in pockets of amended soil between the tree roots.

roots. It's best to dig individual planting holes for your wildflower plants, adding organic matter in these small pockets as needed.

For meadow and prairie gardens, you'll probably be converting an area of your lawn, and you will have to get rid of the sod before you plant. Thorough site clearing is essential, or weeds and unwanted grasses will regrow and choke out your wildflower plants. You can cut and remove the sod, mulch it to death, till it repeatedly, or kill it with an organic weed control like Scythe.

Clearing sod. Removing sod is hard, heavy work. If you're clearing a large site, you may want to rent a gas-powered sod cutter to slice through the grass and matted roots. To clear sod by hand, start by outlining the site with a garden hose or a piece of rope. Cut through the sod with a sharp spade all around the outline. Then use the spade to cut the sod into slices. Next, working with your spade's blade almost parallel to the soil surface, cut under the sod to sever the grass roots. You may find it easiest to work on your knees while doing this part. After cutting up the sod, roll it up and cart it to a corner of your yard to compost.

Use a sharp, flat-bottomed spade to strip sod before planting a meadow or prairie garden.

If the soil beneath the sod is compacted, or if you're going to start the garden from seed, till and rake the soil shallowly. Don't till or plow deeply because you may damage the soil structure and bring up subsoil.

If you can't plant within the first week after clearing the site, annual weeds will undoubtedly sprout. You can prevent this by covering the site with mulch or a tarp until you're ready to plant. Or, you can simply work over the site with a hoe, cutting the seedlings off just below the soil surface. If a large crop of weeds springs up, you can spray them lightly with an organically acceptable weed killer such as Superfast Weed Killer or Scythe.

The main disadvantage of removing sod is that you lose some of the topsoil at the site. If you have enough time available before you plan to plant, turn the sod upside down and leave it to decay on the site. This will retain the topsoil and add additional organic matter as the sod breaks down.

Using mulch. You can save yourself the heavy work of clearing sod if you're willing to plan a full season ahead. Instead of cutting and removing the sod, you can smother it under a heavy layer of mulch.

Start in the spring by mowing the site as close to the ground as possible. Cover the mowed area with a layer of corrugated cardboard or with several layers of newspaper. Then, if you want a rich soil, top the newspaper or cardboard with 8 to 10 inches of organic mulch such as straw, shredded bark, or compost. Let it sit all summer, and replenish the mulch in the fall. By the following spring, the sod will have decomposed, and you can plant right into the thick mulch.

If you don't want such a rich, humusy soil, top the newspapers or cardboard with a heavy, light-excluding cover, like old carpeting or an old shower curtain. Leave this in place the whole season, and then remove it the following spring.

Tilling. You can use your rotary tiller to kill off sod—but be prepared to till more than once. A single pass with the tiller may appear to clear the site, but the grass will undoubtedly resprout. It takes several tillings over time to do the job.

In midsummer, mow the grass as short as you can. Rake up the clippings and use them as mulch in other parts of your yard, or compost them. Till the site, then wait a week or two, until you see seedlings and grasses resprouting. Till again, about 1 inch deep. Wait again, then till again. Repeat tilling every week or two, three or four more times, until very few weeds appear. Remove these by hand, and then your site is ready for wildflowers. (This should work well to prepare a site for fall planting.)

A thick layer of newspaper covered with 8 to 10 inches of compost will kill lawn grass and provide a rich planting bed for woodland wildflowers.

Using organic weed killers. If you feel comfortable about using an organic weed killer, you can prepare a meadow site quickly and inexpensively by spraying it with Superfast Weed Killer or Scythe according to the label instructions. Both of these weed killers are considered acceptable by most organic certification programs. Like insecticidal soaps, they are made from naturally occurring fatty acids. They kill plants by causing the foliage to dehydrate. The spray will kill many kinds of plants, so be careful not to let it get on plants that you *don't* want to kill.

It takes 7 to 14 days for the grass to die after spraying. The weed killers won't kill some perennial weeds, so you'll need to spot-pull weeds that don't die after treatment. You can leave the dead sod in place when you plant, or you can work it into the soil with other amendments before planting. The payoff from using a weed killer is that the dead turf actually serves as a beneficial mulch to control erosion while the new plants or seeds are getting established.

AMENDING THE SOIL

Once your site is clear, you may need to add organic matter such as compost or aged manure to the soil to add nutrients, improve soil structure, and aid in retaining moisture. In general, native wildflowers grow better in less fertile soil than traditional garden plants do.

Most woodland plants require at least 40 percent organic matter in the soil to help hold moisture. For woodland gardens, you'll probably need to add organic matter like shredded leaves or compost to your site to bring it up to this level.

Meadow and prairie plants thrive in relatively poor, often dry soils. In rich soil, the plants may grow too lushly and flop over. Some meadows and prairies, however, are found on deep, rich soils. The easiest way to make a successful meadow or prairie garden is to choose plants that are well suited to the kind of soil at your site. Check "Growing and Propagation" in the encyclopedia entries to find the conditions each wildflower needs.

If you're working with a site where the topsoil was stripped during construction or where the soil is depleted, be sure to add organic matter before you plant any type of wildflower garden.

Adding organic matter. To work organic matter into your soil, start by using a spade or digging fork to loosen the soil. If the site has weeds growing in it, remove them completely, including all spreading roots. Then spread a layer of compost or rotted manure 2 to 4 inches deep over the entire planting bed. Use a fork to break up large lumps of soil and to thoroughly mix the amendments into the top 10 to 12 inches of the soil.

Amending heavy clay soil. If your soil is high in clay, add both organic matter and sand. Loosen the soil first and then spread 2 to 3 inches of compost and 2 inches of sharp builder's sand over the entire bed. Mix these into the top 12 inches of the soil. It's important to add *both* compost and sand. If you add sand alone, you'll probably make the condition of the soil worse.

Once you've worked in the amendments, smooth out the surface of the bed for planting. A metal garden rake is a good tool for this. You should even out dips and ridges in the soil and remove any roots or other debris.

Planting Techniques

You can start wildflower gardens from seeds, small plants, or larger potted plants. For woodland gardens, it's best to start with plants because woodland wildflowers are delicate and slow to establish from seed. The seedlings need coddling indoors or in a cold frame until they have an established crown. For some slow-growing wildflowers, this can take as long as three years!

For meadows or prairies, starting from seeds is much easier and will reduce the cost of a garden. You can also plant a mix of seed and plants, especially if you're planting the garden in place of a lawn and want good-looking results fast. Choose plants of showy perennials like goldenrod, asters, bee balm, and butterfly weed to make a splash while the seedlings are filling in.

WHEN TO PLANT

The peak of the woodland or shade garden is spring. But in order to get a good display, you'll need to plan ahead and plant your woodland garden the previous summer or fall. Plant or transplant early-blooming and ephemeral species—the ones that go dormant after bloom—after they flower. Plant summer- and fall-blooming species in spring or fall. If weather conditions are mild and there's adequate soil moisture, you can plant potted wildflowers any time during the growing season, as long as they won't suffer from high heat or drought right after planting.

It's best to plant prairie and meadow gardens as soon as you clear the site. To plant a meadow or prairie from seed, plan on preparing and planting in spring or fall. Fall is preferable because the seed is fresh and will sprout as soon as spring arrives. Also, in the spring you may have more problems with birds eating the seeds before they germinate and grow. Spring seeding works best on rich, loamy soils that provide ample moisture during the hot summer months.

STARTING FROM SEED

You can plant a small garden from seed by hand, or you can use a fertilizer spreader. If you broadcast by hand, aim for even coverage. Use the seeding rate recommended on the seed mix label, or if you're having a mix custom-made, ask your supplier what the seeding rate should be. In broad terms, the seeding rate will be about 4 1/4 pounds per 1,000 square feet. If you're planning a large-scale garden, you may need to use a large broadcast or drill seeder pulled behind a tractor to spread the seed.

To speed the task of starting a large prairie garden from seed, use a fertilizer spreader to distribute the seeds efficiently and evenly over the planting bed.

After you spread the seed, use a metal rake to rake it into the top inch of soil. If you're only planting a small garden, go ahead and water thoroughly after seeding. But if your garden is half an acre or larger, you'll probably have to wait for rain to moisten your seeds and start the germination process. This won't hurt your garden; prairie and meadow wildflowers and grasses can germinate and grow with relatively sparse amounts of water.

Water your newly seeded meadow or prairie garden deeply and infrequently. These wildflowers and grasses develop deep roots, which help make them naturally drought-tolerant. Daily shallow watering won't encourage them to send their roots deep into the soil.

PLANTING BAREROOT PLANTS

Bareroot plants are just that—plants without soil on their roots. Many mail-order nurseries send plants this way because it's cheaper than sending plants packed with heavy soil around their roots, so you'll pay less in shipping costs. Nurseries grow these plants in fields or in pots. When it's time to ship, one of the nursery employees digs the plants out of the ground or knocks them out of their pots. They shake or wash the soil off the roots and wrap them in spaghnum moss or another absorbent material to keep them moist.

In some cases, nurseries dig plants in advance and store them refrigerated until shipping. This practice usually doesn't hurt the plants, but occasionally it creates problems: Roots may dry out, or if the plants are stored too wet, they may mold or rot. Unpack your plants as soon as they arrive and inspect the roots for signs of damage. If roots are dry, soak then in warm water for several hours before you plant. If they are rotted, cut off the rotten portions. If the plant is done for, make a claim with the nursery. Most nurseries require that you make damage claims immediately.

Planting bareroot wildflowers is fun. (It's also easy if you've prepared your soil correctly ahead of time, as described on page 21.) Start by digging a hole 3 to 5 inches deep and twice as wide as the longest roots of your plant. For example, if the plant has 12-inch roots, make your hole 24 inches in diameter. This rule may seem unreasonable if a plant has very long roots. Don't exhaust yourself on a major excavation; just trim the roots to a manageable length instead.

Make a cone of soil in the center of the hole. Try placing the plant on the soil cone. The plant's crown (the point where the stem joins the roots) should be level with the soil surface. Adjust the cone to the proper height. Then put the plant on it again, and spread the roots evenly around the mound. (You should be able to spread the roots to their full length if you made the hole wide enough.)

When planting bareroot plants, be sure to spread the roots evenly over a mound of soil in the center of the hole. The crown of the plant should be at soil level.

Holding the crown in place, refill the hole with soil. Firm the soil around the crown and top off with more soil. Then spread 2 inches of mulch around the plant, keeping the mulch 1 or 2 inches away from the crown. Water newly planted bareroot wildflowers well. Keep a check on the soil moisture for the first growing season and never let the soil become dry.

The average spacing for bareroot plants and plugs (plants growing in small cell packs) for a meadow or prairie garden is one plant per square foot. Use this spacing for grasses if they form clumps and spread slowly. Leave larger spaces between plants that send out runners, such as mountain mint and whorled milkweed. These plants can spread 2 to 3 feet in just a few growing seasons.

It's best to plant bareroot plants only in spring or fall. If you clear and plant a site in late summer or fall, cover the new planting with a 4- to 6-inch layer of weed-free straw or leaves after the ground freezes.

You can use a bulb planter to make planting holes for bulbous wildflowers like lilies, just as you would for daffodils and tulips.

PLANTING BULBS

Bulbs, corms, tubers, and rhizomes are really bareroot plants too, but they need to be planted differently. (For more about what these root structures are and how to tell them apart, see the Glossary on page 180. Each wildflower's root system is described in the encyclopedia section, so you'll know how to plant it.) For bulbs, dig a narrow hole 2 to 3 times as deep as the bulb is tall, and set the bulb in it. Take special care not to damage the base of the bulb or the new shoot when you refill the hole.

Set the tubers and rhizomes of plants such as trilliums, Virginia bluebells, and bloodroot with the eye (the bud that will sprout) 1 to 2 inches below the surface. Place the horizontal rhizomes of iris at or just below the soil surface.

PLANTING PLUGS

A plug is a small plant grown in cell packs for early transplanting. Plugs are less expensive than larger potted plants. Because plugs are young plants, they are actively growing and establish themselves quickly in the garden. Plugs are easiest to plant and most successful when the soil is well prepared. However, you may even have great success planting them in unprepared soil if you keep the transplant wa-

When you plant an iris, take care to keep the top of the rhizome above the soil line while you spread the roots evenly in the soil.

When planting plugs, break up the rootball before you place it in the hole. Keep the crown of the young plant level with the soil, just as it was growing in the cell pack.

tered until it's well established. If you've killed the sod on your site with an organic weed killer, you can plant plugs directly into the dead sod.

To remove a plug from the cell pack, use your fingers to push the rootball up from the bottom. Inspect the roots to see if any are circling or matted. If so, cut halfway through the rootball with sharp shears and pull the roots apart. Otherwise, just open up the soil ball with your fingers to encourage the roots to spread out. Dig a hole large enough for the plug. If the soil you removed from the hole is hard or forms clods, break it up as well as you can. Set the plug against the side of the hole, keeping the crown level with the top of the hole. Fill in around the rootball, and firm the soil well to hold the plug in place.

Water plants well and keep them moist until the root system is established and the plants resume growth. (This may take an entire growing season.) As long as you can get water to the site, you can plant plugs anytime during the growing season. Spacing is about the same as for bareroot plants (as described on page 25).

PLANTING POTTED PLANTS

Planting a wildflower that's been growing in a pot is not quite as easy as planting a plug. You can't just pop the plant out of the pot and stick it into the ground. It's important to prepare the soil well before planting a potted wildflower. The ease of planting depends on how long the plant has been in the container—plants that have been in the pot too long may need some special treatment.

To remove a plant from its container, cup one hand around the crown of the plant and turn the pot upside down. Shake the container or rap the rim on a solid surface to dislodge the plant. The plant will fall out of the pot into your hand.

Turn the plant upright again and examine the roots. If the plant has only a few circling roots, shake the rootball to remove excess potting soil. This step is important. Most soil mixes used in pots are light and dry out quickly. The soil mix is probably quite different from your garden soil. If you put the plant into the ground as is, the two soils won't work well together. They'll absorb and lose moisture at different rates, which slows down the plant's establishment. Your new wildflower may even die as a result of lack of water, with its roots stuck in a pocket of dried-out soil mix, right in the middle of your well-watered garden. Your plant will fare much better if you shake off that potting mix and then treat it as if it were a bareroot plant (as described on page 25).

If the plant has lots of circling roots, you must disentangle them if you want the plant to adjust quickly and grow well. This step requires a bit of fortitude. Using your fingers, a sharp knife, or pruning shears, attack the roots and free them from bondage. Tease the roots apart with your fingers or cut them where they bend at the bottom of the pot. If the plant has a taproot and it is badly twisted or bent, cut it above the point where it bends. This seems brutal and is hard to do, but the results are worth it. Plants establish themselves much more quickly once their roots are free. After you deal with the circling roots, shake off the potting mix and plant the plant as you would a bareroot plant: Make a mound of soil in the bottom of the

Before planting potted plants, be sure to cut through the circling roots that form a water-resistant barrier between the rootball and the surrounding soil.

planting hole and spread the roots out over the mound. Adjust the height of the mound until the plant's crown is at the same level it was in the pot. Then fill in the hole.

Watering

Pay close attention to watering your wildflower garden for the first season or two after planting, especially if you're growing a woodland garden. Shaded woodland plants grow more slowly than sun-loving plants. Their root systems take one or two full growing seasons to become established after planting. Woodland wildflowers are usually planted under or near trees, where they may have to compete with thirsty tree roots for water. Also, the dense canopy of leaves above can block rain.

If you've started a meadow or prairie garden from seeds or small plugs, you'll also need to check for dryness frequently during the first growing season. You need to keep the soil moist from planting until frost, or the plants won't establish themselves well.

A general rule is that plants need 1 inch of water per week. I recommend that you check the soil frequently to see how moist it is. Different kinds of soil dry out at different rates. Clay soils hold water, while sandy soils dry rapidly. To check your soil, dig down with a trowel at least 2 inches into the soil. The top may look dry even when there is ample moisture beneath the surface. But if the soil is dry 2 inches down, it's time to water.

A soaker hose applies water directly to the soil without losses to wind and evaporation—the usual problem with overhead sprinklers.

When you water, don't just sprinkle plants lightly. Give them a long soaking. Run soaker hoses through the bed, or set up a sprinkler. If you're using a sprinkler, set a rain gauge in the bed to monitor how much water you've applied. It often takes at least two hours of sprinkling to apply 1 inch of water. For soaker hoses, check the packaging, which should give an estimate of how much water the hoses deliver per hour.

Rain will help keep the garden watered, but a summer downpour may not water the garden as much as you'd think. Summer rain often comes down so hard and fast that the water runs off rather than soaking in. Soil can become critically dry in just a few days, especially if it is hot and windy. Always use the trowel test to check your soil!

Once a wildflower garden becomes well established, watering becomes a question of aesthetics rather than survival. In a dry period, some wildflowers and grasses will go dormant to reduce their water needs. If they do, they may not emerge from dormancy later in the season, so you may get reduced late-summer and fall color and display from your garden. If you want to keep the plants growing actively in a drought, give them some supplemental water.

Mulching

Mulch holds moisture in the soil and also blocks some of the sun's heat, which keeps the soil from overheating. It allows the soil to heat gradually in spring, which is important because different kinds of wildflowers and grasses germinate in different temperature ranges. A good blanket of mulch also suppresses weeds and helps protect plants from damage in the winter.

Mulching woodland gardens. The autumnal blaze of leaf color from trees and shrubs is one of the glories of a woodland garden. On the practical side, fall leaf drop is essential to provide winter mulch that protects the soil from rapid temperature change and keeps delicate plant crowns from drying out in winter winds. Leaves decay and release nutrients slowly, so plants have a steady supply of food.

The best mulch for your woodland garden is oak leaves or other durable leaves, either whole or shredded. However, leaves sometimes aren't available when you first plant a garden in spring or early fall. As a substitute, you can use weed-free shredded straw or sifted, shredded hardwood mulch.

A summer mulch reduces weed problems and keeps the soil from drying out too quickly. Mulch also makes the garden look neat.

Woodland gardens need a winter cover of mulch added every year to protect the plants from cold and also to maintain the organic matter content of the soil. Again, leaves are the best choice, followed by weed-free straw (you can put this on straight from the bale in fall, and then rake it off and shred it in the spring).

Mulching meadow and prairie gardens. If you start a meadow or prairie garden from seed, the only mulch you should apply at planting is a light layer of chopped, weed-free straw. If you start with plants or plugs, sifted, shredded hardwood mulch is your best choice. Apply the mulch evenly over the bed 1 to 2 inches deep. Avoid burying the crowns of the new transplants. The mulch will settle and knit together, which helps keep it from washing away.

If you can't get sifted hardwood mulch, use chopped or shredded straw instead. Make sure you use weed-free straw, not hay. Hay still has seedheads, and its seedlings will take over your new garden. Apply an even layer 2 to 3 inches thick. It will settle for an effective 1-inch cover. Mulch is important when you're establishing meadow and prairie gardens, but it's not necessary after the first year or two. Established plants grow so thickly that their spreading stems and leaves are all the mulch they need.

Weeding and Edging

One undeniable fact of gardening is that lawn grasses will always invade adjacent flowerbeds. You can keep grass out by hand-edging your wildflower beds at least twice a year or by installing a solid barrier between planting beds and lawn. Use a sharp, flat spade to cut a neat edge around your bed. If manual edging is too labor-intensive for you, install a high-quality metal edging, called a header, to separate beds from turf. Sink the header about 4 inches into the ground along the full length or width of the bed.

Weeding woodland gardens. Weeds compete with your wildflowers for water and nutrients and may create shade. In general, woodland or shade gardens have few weed problems, especially if you keep them well mulched. You may find tree seedlings popping up as weeds in your

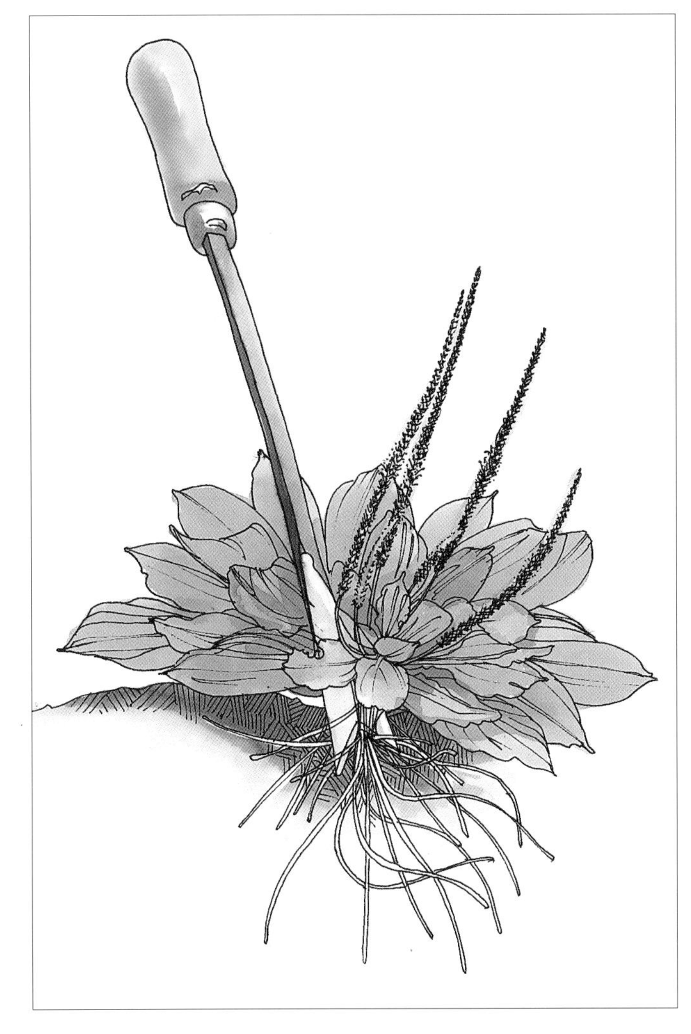

When you weed your wildflower garden, try using an asparagus fork. It works well for both shallow- and deep-rooted weeds.

plantings. Some of these have tenacious roots that are hard to dig out. It may help to grip the stems with a pair of pliers when you pull them. Some of the wildflowers you plant may also spread too enthusiastically. Turn this problem into an opportunity, and trade excess plants with your gardening friends.

Weeding meadow and prairie gardens. In meadow and prairie gardens started from seed, annual weeds will definitely sprout in the first season of growth and will even outgrow your wildflowers. You'll need to mow these gardens once or twice during the growing season to set back the annual weeds and allow your young wildflower plants to compete. In a small garden, you can also weed by hand. If the sod was not completely removed or killed, turf gasses will resprout during the first season and can become a terrible weed problem. Dig them out by hand, and be sure to get out all the rhizomes. Roots of other perennial weeds, especially clover, dandelions, dock, and plantain, may also resprout. Dig these out as well. I find that an asparagus fork or a sharp trowel works well for getting out the deep roots of perennial weeds.

Weeds may or may not be a problem in the second season, depending on the variety and tenacity of weeds present. In the third year and beyond, the grasses and wildflowers will begin to mature and will out-compete or shade the weeds. However, you may still need to hand-weed isolated weeds on occasion.

Cleanup

At the end of the growing season, there are some simple steps you can take to help your wildflower garden make it through the winter.

Woodland gardens. In the fall, after the first hard frost, do a final thorough weeding. Cut all tattered plants to the ground with sharp pruning shears, leaving decorative stems and seedheads standing for winter interest and for birds to eat. Mulch woodland wildflower beds with 2 to 4 inches of leaves or weed-free straw after the ground freezes to help protect the plants over the winter. This is especially important in newly planted areas. Prune trees and shrubs as needed to keep them healthy and vigorous.

In spring, remove the leaves or straw from the crowns of the plants to allow new growth to emerge. After you clear out the bed, if you find that the soil is dry, water thoroughly to get the plants off to a good start.

To make the most of the winter mulch you cleared away, chop it up (you can use a shredder or lawnmower to do this), and then spread it evenly between the plants in your garden. Take care not to bury the crowns. For the growing season, the purpose of the mulch is to conserve soil moisture and keep down weeds, not to protect the wildflowers.

Meadow and prairie gardens. The first two years after planting, your garden may still look a little sparse and tattered by fall. Cut it to the ground by hand or by using your lawnmower. Leave the thatch in place to act as a winter mulch.

Leave the dark, buttonlike seedheads of black-eyed Susan in place to add winter interest and provide seeds for birds.

In subsequent years, when plants are more substantial, wait until spring to cut back dried leaves and stems. They will add winter interest in your yard. If growth is really thick, hand-cut the largest plants and then use a string trimmer or a lawnmower. If you use a string trimmer, don't cut too close to the crown of the plants. Leave 2 inches of stubble above the crowns. If you mow the plants, set the blade 4 inches above soil level and run it slowly over the planting bed. Go over it twice to be sure it is an even cut.

When spring arrives, rake off all winter thatch, as well as any debris that collected over the winter. Don't rake so hard that you remove the hardwood mulch you spread when you originally planted the garden.

For meadow and prairie gardens in rural areas, you may want to try burning off the garden to mimic the natural forces that shape native prairies. Before you burn your garden, you must get a permit from your local municipality or state Department of Natural Resources. Before doing any burning, be sure to read detailed instructions on how to do it safely. Consult "Recommended Reading" on page 184 for references on prairie management. If you don't want to tackle managing a burn yourself, you can hire a prairie restoration company to do the job for you.

Burning mimics the natural cycle of the prairie. It's the fastest way to clean up a large prairie garden to make way for the new season's growth.

The first burn should be in the third or fourth year, repeated every three or four years thereafter. *Don't* set fire to any garden unless you have a valid permit and are sure you know what you're doing.

Pests and Diseases

Native plants are remarkably pest- and disease-free. If you've planted them in a site that meets their needs, you may never need to take any steps to control insect pests or diseases. The beneficial insects that are attracted to your wildflower garden will do their best to keep any pest outbreaks under control.

INSECTS

Two particularly damaging insect pests of wildflowers are borers and aphids. But if you keep an eye out for them and act quickly, they can both be controlled organically.

Borers. Borers are the larvae of moths that tunnel into the rhizomes or crowns of plants such as irises, columbines, and alumroots (*Heuchera* spp.). They eat through the stems from the inside out, which often kills the plant outright. The best method of control is to squash the larvae. However, the borers will often have done severe damage before you realize that the plants are infested. Pull and destroy badly damaged plants. Don't leave them in the garden because the borer eggs can overwinter on old leaves and stems. The following spring, you can take preventive action by dusting the soil and bases of the plants with pyrethrin dust. Pyrethrin is a botanical pesticide derived from pyrethrum daisies and contains several substances that kill insects on contact. Pyrethrin is harmful to some beneficial insects, so don't dust it indiscriminately around your garden.

Aphids. Aphids are easier to spot than borers and are generally easy to control if you take action quickly. Aphids are tiny, pear-shaped, soft-bodied insects that suck plant juices. They attack succulent leaves, stems, and flower buds. They feed on many kinds of wildflowers, including asters, black-eyed Susans, columbines, phlox, milkweeds, and bleeding heart.

Many garden insects are helpful. For example, bees pollinate flowers and praying mantises may eat pest insects.

A small number of aphids is not a cause for alarm. They will be eaten by beneficial insects before they harm your plants. If you have an infestation, one easy way to kill aphids is to squash them by running your fingers up infested stems. You can also control them by spraying infested plants with a stiff spray of water from a hose.

Ladybird beetles are important predators of aphids and other pest insects.

Your wildflower garden will attract beautiful beneficial insects like tiger swallowtail, which feeds on nectar.

DISEASES

Using techniques that prevent disease is more effective than trying to stop a disease outbreak after the fact. That's because there aren't organic sprays that will kill disease organisms that have invaded your plants. To keep your garden disease-free, give plants the growing conditions they need, and clear any diseased plant parts out of your garden as soon as you find them. Rot, rust, and powdery mildew are some common plant diseases that may trouble your wildflower garden.

Bacterial and fungal rot. If your plants turn black at the base, have yellow leaves, or just topple over, they may be suffering from a bacterial or fungal rot. Rot diseases are not too grave a threat to most native wildflowers. The key to prevention is to match your plant to your site. Plants like blackfoot daisy and Maryland golden aster, which are native to naturally dry sites with well-drained, sandy soil, are most likely to suffer from rot. Or, if you have a protracted spell of wet or humid weather, rot problems may appear on plants that usually aren't bothered.

The best treatment for rot problems is to pick off infected leaves and prune off infected stems. Also clear away any diseased debris around the crowns of the plants. Destroy the diseased plant parts by burning them, or dispose of them in sealed containers with your household trash.

Rust. Rust is a fungus which appears as yellow or orange spots, bumps, or pustules on the undersides of leaves. As the spots grow, the upper leaf surface turns pale. Goldenrods, asters, mayapple, and jack-in-the-pulpit are a few of the plants that are susceptible to rust. Most rust species have two host plants. The fungus lives out different parts of its life cycle on the two different hosts. For example, the rust fungus that attacks goldenrod and asters infects pine tree needles in the spring and early summer and overwinters on the wildflowers.

The spots start off small, but as the rust grows, the leaves may be disfigured or destroyed. Remove and destroy infected leaves as soon as you spot them. In severe cases, you'll have to pull up and destroy entire plants. Sulfur is an organically acceptable fungicide that can help prevent the spread of rust. If you have plants that have had rust problems in previous seasons, begin spraying with sulfur before symptoms appear, and continue applications every seven to ten days when conditions are wet or humid. Don't spray sulfur when temperatures are above 80°F, or it may damage the plants.

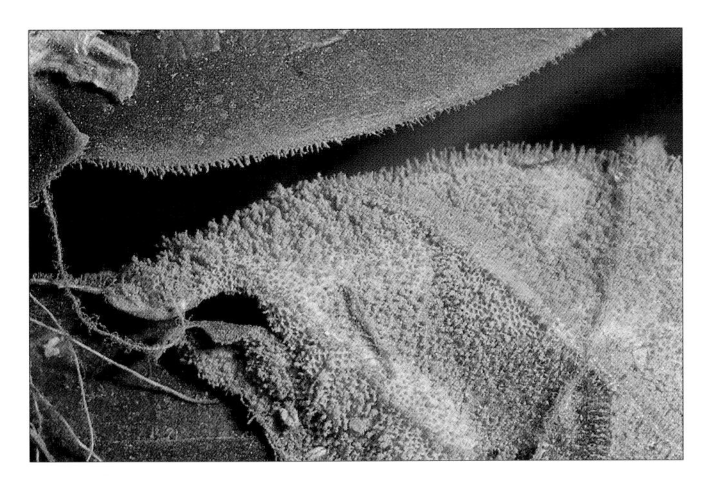

This close-up view shows the orangy leaf coating that is typical of rust disease.

Powdery mildew. Powdery mildew is a fungus that affects foliage and occasionally stems. It transforms the surfaces of leaves from green to white, looking like a coating of plaster dust. Powdery mildew affects asters, green and gold, bee balm, phlox, and many other wildflowers.

Often, a prolonged dry spell followed by humid conditions or a rainy spell brings on powdery mildew symptoms. Keeping plants well watered helps prevent powdery mildew, as does good air circulation. If crowns of susceptible plants are dense, thin the stems by one-third to open up the clump. You can spray susceptible plant once a week with wettable sulfur as soon as the first growth appears, but prevention is more successful than control. Clean up infected debris from the garden and remove and dispose of infected stalks in the autumn so spores don't overwinter.

ANIMAL PESTS

As much as we love animals in our gardens, sometimes they become pests. They eat or dig up our plants and generally wreak havoc. Tolerance is often the best approach. Voles, squirrels, rabbits, woodchucks, and deer are the main pests of wildflower gardens.

Voles. Voles eat the roots, crowns, and young shoots of plants. You may never see voles in your garden because they work under cover of soil or mulch. If an otherwise healthy plant suddenly wilts and collapses, give it a tug. If it comes up freely and rootless, you've got voles. They also work under the snow in winter, eating the dormant crowns until spring. The best way to control voles is to remove as much excess mulch and debris from the garden as you can. You may also make the choice to trap the voles. If you can locate the burrows or the runs under mulch or snow, place rat traps baited with carrot or apple where they are likely to encounter it. Keeping a cat or dog around your yard can also deter voles.

Squirrels, rabbits, and woodchucks. For squirrels, rabbits, and woodchucks, try using repellents. Human hair sometimes repels these animals (you can collect it from a barber shop or beauty salon). Cayenne pepper or dried blood (available at garden centers) sprinkled around your plants is also a good deterrent. Don't sprinkle cayenne pepper on windy days, as it can severely irritate your eyes.

Deer. Repellents may also deter deer, but they may not be effective in areas with large deer populations. For a large garden, the only effective way to keep deer from eating plants is to put up a fence. Unfortunately, most fences designed to keep deer out are expensive to install, but this is the best long-term solution.

Voles are voracious eaters of roots and young shoots. They can seriously damage plants, particularly during the winter.

Working with Your Garden over Time

As your wildflower garden grows and develops, it will take on a mature look that adds fullness and beauty to your landscape. You'll still need to intervene occasionally, though. Plants will grow too large and need division. Seedlings will come up in the oddest places. These volunteers can enhance the planting—just transplant any that find an inappropriate home. As conditions change in the garden, the plant composition will also change. Aging trees produce more dense shade. A dead tree will open a sunny spot. Enjoy these opportunities that nature provides.

You may decide that a plant doesn't look right where you planted it. Don't worry—if you don't like the plant, dig it up and trade it with a gardening friend for something you like better. If you do like it, find a place where it would look better, then move it. If you decide to transplant wildflowers or grasses, your goal is to move the plant without disturbing it. Try to dig around the entire root system, leaving the soil intact. This requires some knowledge of the plant's root system—whether or not it has a taproot, for example. (You can refer to the encyclopedia entries in this book for information on root systems of specific plants.) Once you've worked the whole rootball loose, insert a shovel or trowel and lift the freed rootball out of its hole. Dig a hole the same size as the rootball at the plant's new location, saving the soil you remove in a bucket. Place the rootball in the hole, press it firmly into place, and mulch. Then dump the soil in the bucket into the original hole.

Propagating Wildflowers

Popularity has proved a problem for our native wildflowers. To meet increased demand for plants, some suppliers and gardeners turn to collecting plants from the wild. My message to you is simple: Don't collect plants from the wild. Even if your intentions are good, you may disturb sensitive ecosystems. You can also dig up problems hidden in the soil around wild-collected plants. The soil may contain weed seeds or disease organisms that will plague your gardens. Wild collecting is not necessary because we now know how to propagate most wildflowers.

Most native species are easy to propagate from seeds, cuttings, and divisions. All it takes to propagate native wildflowers is an understanding of the requirements of the individual species. Starting seeds, taking stem or root cuttings, making divisions, and layering are established techniques that work as well for wildflowers as they do for common garden perennials and other plants.

The fleshy red fruits of Jack-in-the-pulpit contain one large seed, which should be stored in a cool, moist place until time for sowing.

STARTING SEEDS

Although some dealers offer native plant seeds, often you must collect seeds of the species you wish to increase from plants you've already bought for your garden. This is especially true of the spring-blooming woodland ephemerals that go dormant after flowering. Their seeds must be sown immediately after harvest, and these seeds are not available commercially. Seeds of most woodland ephemerals ripen 4 to 8 weeks after flowering. Some species, such as trillium, bloom in April but their seeds do not ripen until midJune.

Collecting seeds. It's best to harvest wildflower seeds as soon as they ripen. The seeds may be borne inside pods or berries. Pod swelling is a sign of ripening. When pods or berries change color, it often means that seeds are mature.

Check plants frequently as maturity approaches, or the seeds may disappear before you get to them. The seeds of many spring bloomers have fleshy, winglike structures that are attractive to ants. Ants will carry the seeds away as soon as the fruit is ripe. Pods of other plants eject ripe seeds with explosive force. To avoid losing the seeds from these plants, enclose the fruit in a mesh net. (You can use a section of panty hose tied with twist ties for this.)

Timing is less important with wildflowers whose seeds remain attached after ripening. Learning the habits of the plants and keeping a close watch are the best ways to ensure successful seed collection.

Sorting seeds. It's best not to try to store seeds of spring ephemerals such as spring beauty, trout lily, and bloodroot. Instead, sow them outdoors immediately after you gather them.

If you want to store other types of seeds, clean them thoroughly first. Wash all pulp from berries and remove debris from capsules and pods. The ideal temperature for seed storage is 35 to 40°F, so storing them in your refrigerator is fine. The condition of the ripe seed will tell you how to store it.

It's easy to store seeds that ripen dry in their pod or capsule. Simply put the seeds into small manila envelopes or glassine envelopes (like the kind used for postage stamps), label them, and seal them.

Seeds from fleshy fruits need moist storage. Your goal is to create a high-humidity environment without having the seeds sitting in water, which can lead to rot. I like to use small sealable plastic bags and whole-fiber sphagnum moss. Wet the moss thoroughly and then wring it dry. Let it sit a minute, and then wring it out again so that it doesn't drip at all. Then make a nest of the moss in one hand, and put the seeds on the moss. Grab another handful of moss with your other hand, and put it on top of the seeds. Then carefully slide the moss-and-seed sandwich into a plastic bag, and seal it two-thirds of the way. Starting at the bottom of the bag, run your hand over it to push air out and then seal it the rest of the way. Don't forget to label the bag.

Sowing seeds. You can sow wildflower seeds directly in outdoor beds right after collecting them. If you decide to clean and store them, start your seeds in early spring in commercial flats or clean, recycled containers, and place them on a windowsill or in a cold frame or greenhouse.

The main advantage of starting seed indoors is that you can control environmental conditions. When planted outdoors, your seeds or seedlings may be ruined by soil diseases or insects, or weather conditions may not be good for germination. Indoors, you can take special care to get the best germination rates and produce healthy seedlings. It's easier to keep track of the seedlings, too.

Store dry seeds in envelopes and moist seeds in plastic bags. When it's time to plant, sow the seeds evenly over the soil surface. Label the flats with the name of the plant and the date it was sown.

Timing of sowing seeds can be tricky. You'll need to calculate backward from the date that the plants can be set out in the garden to decide when to start your seeds. Allow one month for the plants to grow in flats, four to six weeks for seeds to germinate and seedlings to emerge, and another four to six weeks for seeds that need some kind of treatment to break dormancy. (Refer to the encyclopedia entries in this book for information about dormancy of specific plants.)

Be sure that all containers you use for starting seeds and growing transplants have drainage holes in the bottom. Fill flats with a 1:1 mix of quartz sand and milled sphagnum moss. Spread seeds evenly over the surface of the mix, then cover them to a depth equal to the seeds' diameter. For seeds of wildflowers that need very good drainage, cover the seeds with pure sand. For wildflowers that tolerate average soil conditions, it's okay to use more seed mix to cover them. Leave very fine seeds uncovered.

Treating seeds. Many seeds need a cold or warm, moist treatment (called stratification) to ripen fully or to overcome dormancy. Sow seeds that need to be stratified in flats, then water the flats and let them drain. Seal them inside plastic bags and store them in a refrigerator or other cool place. For most wildflower seeds, four weeks of cold treatment will be enough to overcome dormancy. But to be extra sure, store them in the cold for six weeks. Do check on the seeds after four weeks because some may start to germinate. If you find germinating seeds, remove the flat from the cold, and move it to your growing area. From there on, treat them the same as freshly sown seeds. Germination time varies with species. (You'll find information about germination time of specific wildflowers in the encyclopedia entries.)

Transplanting your seedlings. Transplant seedlings into small pots or plastic cell packs containing a standard soil mix once the first true leaves develop. After transplanting, put the flats in a sunny window or under lights and water regularly, so the soil stays evenly moist. Fertilize with a half-strength solution of fish emulsion as needed.

A vigorous young plant that is ready for the garden has a well-developed crown, healthy leaves, and roots that fill the cell pack but are not tangled.

You'll probably need to keep the seedlings growing in flats for at least one month before you can plant them out in the garden. To check whether plants growing in cell packs are ready for planting out, pop one plant out of its cell. The roots should be growing throughout the mix. The plant should also have a good rosette or clump of leaves. If the plants aren't big enough after one month, and if the weather has warmed sufficiently, you can move the flats outside while the plants continue growing.

TAKING CUTTINGS

Growing wildflowers from cuttings is an easy way to produce blooming-size plants in a short time. Many perennials such as green and gold and phlox root quickly from cuttings. Timing is important when taking cuttings; if you take them too early or too late, they won't root. The best time for taking cuttings is midMay to midJuly, after growth hardens. Rooting is most uniform in this period, and perennial wildflowers will produce the best flowering the following spring.

Taking stem cuttings. To take cuttings, use a sharp knife to cut through a stem at an angle, just below a spot where a leaf is attached (this spot is called a node). Make cuttings 3 to 6 inches long from the growing point. With some wildflowers that root easily and produce long stems, such as asters or tall phloxes, you can also make cuttings from the parts of the stem below the growing point. These cuttings are called medial cuttings.

Strip the leaves from the lower half of each cutting. Take care not to damage the stem. If the foliage is large, remove half of each of the remaining leaves.

Push the stripped ends of cuttings into flats or pots filled with a mixture of equal parts of peat moss and perlite. Protect the flats from drying out by covering them with plastic or putting them in a cold frame or (if you're lucky enough to have one) a greenhouse mist bed. You can even improvise a rooting chamber by setting uncovered flats in an old aquarium and covering the aquarium with a pane of glass or a piece of plastic. Check the flats frequently for moistness. Never let the rooting medium get dry.

Taking tip cuttings is a fast, easy way to produce blooming-size plants in a single season.

You can grow new plants from root cuttings taken from plants with fleshy, fibrous root systems. This is the fastest method for propagating some wildflowers, including shooting stars (Dodecatheon spp.).

Cuttings will root in two to four weeks. Harden them off by gradually reducing moisture before transplanting them into the garden or containers filled with soil mix.

Taking root cuttings. You can also produce new plants of some species from sections of their roots. Try taking root cuttings from butterfly weed, phlox, purple coneflowers, and wild ginger in late winter; they should grow easily and quickly. Dig down beside plants in the garden and cut off sections of healthy roots. Then cut these up into 2- to 4-inch-long sections.

Clean, sharp builder's sand or quartz sand is the best medium for root cuttings. When you put the root cuttings in the sand, be sure that the end that was closest to the plant is oriented upward. Cuttings planted upside down will not root. Transplant when shoots are 3 inches tall.

DIVISION

Division is the easiest method of propagation. Plants with multiple crowns and those that form clumps with short runners or rhizomes such as penstemons, phlox, wild gingers, and bloodroot are the most likely to be successful after division. Lift clumps in early spring or after flowering, remove the excess soil, and pull or cut the crowns apart. You may be able to do this with a small knife or with pruning shears. For plants with very tough crowns such as goat's beard or mature phlox, try using a saw or hatchet to cut sections apart. Take care to leave enough roots on each new plant to ensure survival. If some of the divisions have skimpy roots, plant them in a special holding bed, where you take extra care of them until they're well rooted.

Part II

Encyclopedia of Wildflowers

◀ *Most of us think of asters as fall-blooming plants for sunny sites like wildflower meadows and perennial borders. But this species thrives with ferns in a woodland setting. You'll find wildflowers for every situation in the pages that follow.*

Actaea pachypoda

White Baneberry

Pronunciation	ak-TEE-uh pak-ee-POH-duh
Family	Ranunculaceae, Buttercup Family
USDA Hardiness Zones	3 to 8
Native Habitat and Range	Deciduous and mixed coniferous woodlands in rich, neutral to acidic soils from Quebec to Ontario, south to the Georgia uplands and Oklahoma

White Baneberry

DESCRIPTION

White baneberry is a bushy woodland plant noted for its showy autumn fruits. Fuzzy, white, spring flowers bloom in dense, rounded spikes on sturdy stems. The flowers lack petals, so the floral display comes from many short, broad stamens. Oval, white, ¼-inch fruits on showy, red stalks follow the flowers. Fruiting spikes may grow to 1 foot long. The end of each fruit bears a deep blue spot; because of this, white baneberry is also called "doll's-eyes." Birds eat the berries, which are poisonous to people. Plants grow 2 to 4 feet tall and have sharply toothed, fernlike compound foliage. Mature clumps have many stems arising from woody crowns and may reach 3 feet across.

GARDEN USES

A mature clump of white baneberry in full fruit creates a striking display in the autumn garden. Use a single plant for dramatic accent at the edge of the path in a woodland garden or as a focal point at the end of a view. Mass plantings are effective among ferns and foliage plants such as wild ginger (*Asarum* spp.), interrupted fern (*Osmunda claytoniana*), wood ferns (*Dryopteris* spp.), and shield ferns (*Polystichum* spp.) as well as amid the flowering spikes of cardinal flowers and blue lobelias (*Lobelia* spp.), white wood aster (*Aster divaricatus*), monkshoods (*Aconitum* spp.), and gentians (*Gentiana* spp.). Combine flowering baneberries with columbines (*Aquilegia* spp.), wild bleeding heart (*Dicentra eximia*), trilliums, and bellworts (*Uvularia* spp.).

GROWING AND PROPAGATION

Plant baneberries in moist, humus-rich soil in partial to full shade. Baneberries are tough, long-lived perennials that spread slowly from thick crowns to form showy clumps. Plants seldom need division. Enrich the soil with shredded leaves or compost to keep plants healthy. Remove the pulp from around the several shiny, brown seeds found in each berry, and sow seeds when they are fresh. Baneberry seeds have a complex dormancy that requires several treatments to overcome. Give them 3 weeks of warm, moist stratification, followed by 5 to 6 weeks of cold, moist stratification. Seedlings are slow to develop. Transplant to the garden the second spring. Self-sown seedlings will appear.

A RED-FRUITED *ACTAEA* SPECIES

Red baneberry (*Actaea rubra*) is similar to white baneberry but is more northern in distribution. The leaflets are deeper green and broader, the flowers and berries are borne in tighter clusters, and the berries are a deep, glossy red. Plants require cool woodland conditions and acidic soils. They are intolerant of excessive summer heat and limy soils. The form *neglecta* has ivory berries and is found from Labrador to Alaska, south to New Jersey and California. Zones 2 to 8.

Allium cernuum

Nodding Onion

Pronunciation	AL-ee-um SIR-new-um
Family	Liliaceae, Lily Family
USDA Hardiness Zones	3 to 9
Native Habitat and Range	Open, rocky woods, meadows, and prairies in a variety of soils from New York to British Columbia, south in the mountains to Georgia and Arizona

Nodding Onion

DESCRIPTION

Nodding, tear-shaped buds open into loose, drooping clusters of starry, pink flowers in midsummer. The flowerstalk stands 1½ to 2 feet tall. The straplike, pale green leaves are 1 to 1½ feet long and form dense clumps up to 1 foot across. Foliage remains attractive all summer, and the papery, dried seedheads are decorative in autumn.

GARDEN USES

For a stunning combination of forms and textures, plant drifts of nodding onions with a foreground planting of wild petunia (*Petunia parviflora*) and a backdrop of rattlesnake master (*Eryngium yuccifolium*), purple coneflower (*Echinacea purpurea*), and switch grass (*Panicum virgatum*). Try nodding onion's unique, globe-headed form in formal borders, meadows, and prairie gardens. Nodding onion grows well in containers, too.

GROWING AND PROPAGATION

Nodding onion thrives in average to rich, well-drained soils in full sun or light shade. In rich soils, clumps grow quickly and become quite dense, with many flowering stems. Eventually, flowering decreases due to overcrowding. Divide crowded clumps in early spring or as they go dormant. The individual bulbs are easily pulled apart. Replant the divisions in amended soil. They are easily grown from seed sown outdoors when ripe. Self-sown seedlings will appear.

ANOTHER *ALLIUM* SPECIES FOR ROCK GARDENS AND MEADOWS

Prairie onion (*Allium stellatum*) is often mistaken for nodding onion, but the pale pink to rose flowers are smaller and more open and are carried in domed upward-facing clusters. The open clumps have a wispy, delicate appearance. Plants grow 6 to 12 inches tall; the narrow, flattened leaves are 1 to 1½ feet long. Plants vary in height and vigor depending on the moisture and fertility of the soil. Plant in gravelly to rich, well-drained soil in full sun or light shade. Found in prairies, savannas, and open woods and on rocky slopes from Ontario and Saskatchewan, south to Illinois and Montana. Zones 3 to 8.

Amsonia hubrectii

Arkansas Blue Star

Pronunciation	am-SOH-nee-uh hue-BRECK-tee-eye
Family	Apocynaceae, Dogbane Family
USDA Hardiness Zones	4 to 9
Native Habitat and Range	Gravel bars, creek beds, streamsides, and bottomlands in west-central Arkansas and adjoining Oklahoma

Arkansas Blue Star

DESCRIPTION

Arkansas blue star forms billowing clumps of fine-textured, narrow, 3- to 4-inch foliage topped in spring with small domed clusters of starry, five-petaled, ½-inch blue flowers. The summer fruits resemble green cigars that darken to brown as the seeds ripen. Mature clumps provide a soft, 3-foot, shrublike presence throughout the summer. In fall the foliage turns a striking peach to golden color.

GARDEN USES

The fine texture of the Arkansas blue star's foliage makes it perfect for combining with bold flowers such as hibiscus and solid masses of flowers such as yarrows. The plant is at its best in the fall, when it adds its foliar display to the show put on by the season's last aster and mum flowers. Use Arkansas blue star singly as a specimen or in sweeping drifts for a bold, dramatic effect.

GROWING AND PROPAGATION

Plant Arkansas blue star in average to rich, moist but well-drained soil in full sun or partial shade. Give these large plants plenty of room. Set clumps at least 2 feet apart. Established plants are drought-tolerant. Cut plants growing in shade back to 10- to 12-inch stems after flowering. The new growth will be compact and the plants may reflower. Although plants are southern in distribution, they show remarkable hardiness outside their native range. In colder zones, a winter mulch is advisable. Divide overgrown clumps in autumn. Propagate from tip cuttings taken in early summer or by seed. Collect and store seeds in the refrigerator until ready to sow. Soak them overnight in warm water before sowing. Sow ripe seed outdoors for germination the following spring. Self-sown seedlings will appear.

ANOTHER NARROW-LEAVED *AMSONIA* SPECIES

Downy blue star (*Amsonia ciliata*) is a fine-textured, bushy plant that resembles Arkansas blue star but is more delicate. Its wiry stems reach 1½ feet, and the narrow, downy, 2- to 3-inch leaves cluster toward the end of the stems. The small, pale blue flowers are less showy than other species, but the foliage is lovely all season and turns golden orange in the fall. Plant in average to rich, moist soil in full sun or partial shade. Found in open woods and prairies and on roadsides and slopes, often on limy soils, from North Carolina and Florida, west to Texas. Zones 4 to 10.

Amsonia tabernaemontana

Willow Blue Star

Pronunciation	am-SOH-nee-uh tab-er-nay-mahn-TAN-uh
Family	Apocynaceae, Dogbane Family
USDA Hardiness Zones	3 to 9
Native Habitat and Range	Moist, loamy, or sandy open woods, roadsides, and meadows from New Jersey to Illinois, south to Georgia and Texas

DESCRIPTION

Willow blue star is a tough plant with steel blue flowers borne in terminal clusters in spring. The bright green, 4- to 6-inch lance-shaped leaves remain attractive all season, giving a shrublike appearance to the dense 3- to 3½-foot clumps. The foliage turns yellow to fiery orange in the fall. Leaf width varies among individual plants. The variety *salicifolia* has narrow, deep green, glossy leaves.

GARDEN USES

Willow blue star adds structure to the garden. Use its broad rounded form to contrast with tall plants such as meadow rues (*Thalictrum* spp.), Joe-Pye weeds (*Eupatorium* spp.), asters, and sunflowers. A mass planting makes a good hedge or low screen. In beds and borders, combine the flowers with the showy foliage of lady's-mantle (*Alchemilla mollis*), lamb's-ears (*Stachys byzantina*), and ferns or with colorful cranesbills (*Geranium* spp.), bellflowers (*Campanula* spp.), and Siberian iris (*Iris sibirica*).

GROWING AND PROPAGATION

Willow blue star is an easy-care plant that thrives on neglect. Grow in average to rich, moist soil in full sun or partial shade. Willow blue star does not flower well in dense shade. Mature plants attain gigantic proportions. Keep their size manageable by shearing them down to 6 to 8 inches after flowering. This also promotes compact growth, making plants less apt to flop in wind and rain. For propagation information, see *Amsonia hubrectii*. Self-sown seedlings may be plentiful.

Willow Blue Star

OTHER RECOMMENDED *AMSONIA* SPECIES

Blue star (*Amsonia illustris*) is similar to willow blue star but the leaves are narrowly lance-shaped, lustrous, and leathery. The flowers are slightly smaller, and the plants are more compact. Plant in moist soil in full sun or light shade. Found in open low woods and wet prairies and at streamsides from Missouri and Kansas, south to Oklahoma and Texas. Zones 4 to 9.

Dwarf willow blue star (*Amsonia tabernaemontana* var. *montana*) is a compact plant that grows to 1½ feet tall with proportionately larger clusters of waxy flowers. This diminutive variety is especially useful in small gardens or in tight places where economy of size is essential. Zones 4 to 9.

Louisiana blue star (*Amsonia ludoviciana*) forms large open clumps to 3 feet or more with an equal spread. The broadly lance-shaped, dull green leaves have fuzzy undersides and are sparsely arrayed on the stems. The foliage turns burnt orange in fall. Plant in average to rich, dry to moist soil in full sun or partial shade. Plants are adaptable and easy to grow well beyond their natural range in the deciduous woodland borders and coastal savannas of Louisiana. Zones 6 to 9.

Anemone canadensis

Canada Anemone, Meadow Anemone

Pronunciation	uh-NEM-oh-nee can-uh-DEN-sis
Family	Ranunculaceae, Buttercup Family
USDA Hardiness Zones	2 to 8
Native Habitat and Range	Low woods, wet prairies, ditches, and marshes in a wide variety of soils from Quebec to Alberta, south to Maryland and New Mexico

Canada Anemone

DESCRIPTION

Canada anemone brightens early summer with 2-inch bright white flowers held on slender stems above a whorl of three deeply cut leaves. Plants grow 1 to 2 feet tall from fast-growing rhizomes. Its vast colonies of basal foliage and flowerstalks create a spectacular display when in bloom.

GARDEN USES

Plant exuberant Canada anemones in the meadow or in the moist soil of a bog garden. They are excellent groundcovers and are perfect for planting beneath shrubs. The snow white flowers add grace to spring combinations of bulbs, astilbes, phlox, old-fashioned bleeding heart (*Dicentra spectabilis*), and primroses. In the wild garden, combine them with false Solomon's seal (*Smilacina racemosa*), Virginia bluebells (*Mertensia virginica*), ferns, and columbines (*Aquilegia* spp.).

GROWING AND PROPAGATION

Give Canada anemones rich, evenly moist soil in sun or light shade. They spread by creeping underground stems to form showy clumps and may become invasive. They are best used in the wild garden or planted in bottomless containers to control their spread. Canada anemones are easy to propagate from root cuttings taken after the plants are dormant or by sowing fresh seed outdoors.

OTHER *ANEMONE* SPECIES FOR MEADOWS AND PRAIRIES

Carolina anemone (*Anemone caroliniana*) has green, pale blue, or white flowers with rounded to slightly pointed petals. Plants bear soft-hairy, three-lobed leaves in a whorl halfway up the 1-foot-tall bloom stalks. Plant them in average to rich, moist soil in full sun or partial shade. Found in meadows and woodland margins from North Carolina and Tennessee, south to Georgia and Mississippi. Zones 6 to 9.

Thimbleweed or **candle anemone** (*Anemone cylindrica*) is an endearing prairie plant with multiple ¾-inch greenish white flowers held singly on long, erect stalks. Plants grow to 2 feet tall from fleshy-rooted crowns. Below the flowers, a whorl of three deeply dissected leaves graces the stem. The erect, green, cone-like seedheads explode into a cottony froth when ripe. Grow in average to rich, moist to dry soil in full sun or light shade. Found in meadows, prairies, and dry savannas from Maine to British Columbia, south to New Jersey and Arizona. Zones 3 to 8.

Virginia anemone or **thimbleweed** (*Anemone virginiana*) is similar to candle anemone in most respects. The greenish flowers are occasionally white and have pointed, petallike sepals. Give plants average to rich, moist soil in full to partial sun. Found in open woods, clearings, meadows, and roadsides from Nova Scotia and British Columbia, south to Georgia and Kansas. Zones 3 to 8.

Anemone quinquefolia

Wood Anemone

Pronunciation	uh-NEM-oh-nee kwin-kwe-FOE-lee-uh
Family	Ranunculaceae, Buttercup Family
USDA Hardiness Zones	3 to 8
Native Habitat and Range	Deciduous or mixed coniferous woodlands, clearings, and roadsides in moist, rich, limy, or acidic soils from Quebec to Manitoba, south to New Jersey, and in the mountains to Georgia and Iowa

Wood Anemone

DESCRIPTION

Wood anemone is a delicate, spring-blooming wildflower that forms patches of starry, snow white flowers that dance above the forest floor. The ½- to 1-inch flowers are set off by a whorl of three dark green, palmately divided leaves with five leaflets each. Small plants occasionally have only three leaflets per leaf, and some never develop five leaflets. Plants grow 6 to 12 inches tall from slender, fleshy rhizomes to form vast colonies.

GARDEN USES

Wood anemone is one of the few plants that can add the illusion of snow to the spring woodland. Plant them in masses under the shade of trees or in combination with spring bulbs, wildflowers, and ferns. The airy plants will creep among other garden denizens and pop up where space permits. Combine them with merrybells (*Uvularia* spp.), bloodroot (*Sanguinaria canadensis*), hepaticas (*Hepatica* spp.), and sedges (*Carex* spp.). In acidic soils, galax (*Galax rotundifolia*), shortia (*Shortia* spp.), Canada mayflower (*Maianthemum canadense*), and bluebead lilies (*Clintonia* spp.) make excellent companions. Plants go dormant by early summer. Take care not to dig into the dormant clumps when doing summer or autumn planting.

GROWING AND PROPAGATION

Plant wood anemones in moist, humus-rich soil in partial shade. They tolerate a broad range of soil pH. New plants are slow to establish and may take a few years to spread. Plants will go completely dormant after flowering. Plant the rhizomes in early spring or fall. Divide clumps after flowering or when dormant. Sow fresh seeds outdoors.

ANOTHER WOODLAND ANEMONE

Mountain anemone (*Anemone lancifolia*) is similar in most respects to *Anemone quinquefolia*, but its larger leaves have just three leaflets each. The flowers are slightly larger, to 1 inch. Plants grow 10 to 14 inches tall. Plant in moist, humus-rich soil in partial to full shade. Found in deciduous woodlands, on roadsides, and in meadows of the Piedmont and lower mountains from Pennsylvania, south to Georgia. Zones 5 to 8.

Anemonella thalictroides

Rue Anemone

Pronunciation	uh-nem-oh-NEL-uh thuh-lik-TROY-deez
Family	Ranunculaceae, Buttercup Family
USDA Hardiness Zones	3 to 8
Native Habitat and Range	Open, rocky woods, clearings, and roadsides in a variety of soils from New Hampshire to Minnesota, south to Florida and Kansas

DESCRIPTION

Rue amemones are among the most delicate of spring wildflowers. Their thin, wiry stems sway with ease in the slightest breeze. The exquisite flowers have five to eight petallike sepals in a cluster atop the stem. Petals vary from pure white to shell pink and rose, and a deep pink-flowered plant may grow alongside one that bears pure white flowers. Plants produce compact clumps of flowering stems from clusters of fleshy tubers. Each stem is whorled with compound leaves with smooth, rounded leaflets that have several blunt lobes. Plants go dormant by midsummer.

GARDEN USES

Give captivating rue anemones a special place in your garden. Combine them with delicate ferns and low groundcovers such as wild gingers (*Asarum* spp.), hepaticas (*Hepatica* spp.), and Allegheny pachysandra (*Pachysandra procumbens*), which will not swamp them while in bloom but will fill the void left when they go dormant. Spring bulbs, primroses, violets, lungworts (*Pulmonaria* spp.), and epimediums (*Epimedium* spp.) are excellent companions.

GROWING AND PROPAGATION

Plant rue anemone in rich, moist but well-drained soil in light to full shade. Plants in sunny locations flower more freely and remain in active growth longer. If the soil becomes dry, plants will go dormant but will not be harmed. Avoid unnecessary competition from larger plants as they will surely win, and the rue anemone will disappear. Divide the tubers as the plants go dormant, or sow fresh seeds outdoors. Seedlings will appear the following spring. Seedlings take several years to flower.

Rue Anemone

SELECT CULTIVARS OF RUE ANEMONE

'Betty Blake' sports soft, starry, lime green, double flowers with narrow petals. Plants bloom for several months. Also called 'Green Dragon'.

'Cameo' bears lovely, soft pink, double flowers and is very vigorous. Some of the outer petals are elongated, giving a mild starburst look.

'Schoaf's Double Pink' has deep rose-pink pompoms that last for a week or more. It is also a robust grower.

Aquilegia canadensis

Wild Columbine

Pronunciation	ack-wih-LEE-gee-uh can-uh-DEN-sis
Family	Ranunculaceae, Buttercup Family
USDA Hardiness Zones	3 to 8
Native Habitat and Range	Open, rocky woods, rock outcroppings, savannas, and roadsides in near-neutral or occasionally acidic soils; naturalized in a variety of habitats from Nova Scotia to Saskatchewan, south to Florida and Texas

Wild Columbine

DESCRIPTION

Airy clusters of red and yellow flowers nod atop slender stems, blooming for 4 to 6 weeks in spring and early summer. Each flower consists of five long-spurred petals and five petallike sepals surrounding a central column of projecting yellow stamens. Wild columbines produce lush mounds of compound foliage with fan-shaped leaflets. The foliage remains attractive all season and persists through the winter in mild zones. Plants grow from a thick taproot to reach 1 to 3 feet tall. The selection 'Corbett' has pure yellow flowers.

GARDEN USES

Plant columbines in clusters or large sweeps to complement spring wildflowers, perennials, and late bulbs. In beds and borders, they start blooming with the late tulips and continue through the early summer when cranesbills (*Geranium* spp.), irises, peonies, and lupines (*Lupinus* spp.) bloom. Use columbines around the base of tall, shrubby perennials such as blue stars (*Amsonia* spp.) and baptisias (*Baptisia* spp.). In the rock or woodland garden, they are elegant with Virginia bluebells (*Mertensia virginica*), trilliums, wild blue phlox (*Phlox divaricata*), and ferns. Accent a shrub planting with casual drifts of columbines in sun or shade.

GROWING AND PROPAGATION

Plant columbines in light, average to rich, moist but well-drained soil. Heavy or soggy soils will shorten their lives. They grow well in full sun or partial shade. Cut flowerstalks to the ground after flowering or seed-set to promote new foliar growth. Columbine foliage is often attacked by leafminers. They create pale, tan tunnels or blotches between the upper and lower leaf surfaces. Remove and destroy affected leaves immediately and continually. Spray weekly with insecticidal soap. Borers also may be a problem and can cause whole plants to collapse dramatically. Remove and destroy infested plants. Sprays of *Bacillus thuringiensis* (BT) may be effective if applied at the first sign of attack. Plants self-sow freely, but only the true species produce attractive seedlings. Hybrid seedlings are inferior. Sown outdoors, seed germinates easily in fall or spring. Sow seeds indoors after 4 weeks of cold storage.

Aralia racemosa

Spikenard

Pronunciation	uh-RAY-lee-uh ray-sih-MOW-suh
Family	Araliaceae, Ginseng Family
USDA Hardiness Zones	3 to 7
Native Habitat and Range	Rich deciduous or mixed coniferous woods in humusy, neutral to acidic soil in partial to deep shade from New Brunswick to South Dakota, south to North Carolina and New Mexico

Spikenard

DESCRIPTION

Although herbaceous, mature spikenards are commanding plants that grow to 6 feet tall and resemble large shrubs. The huge, 2½-foot leaves are twice pinnately divided. Individual heart-shaped leaflets are 4 to 6 inches long and coarsely toothed. The small, spherical flower clusters are gathered into 1- to 3-foot terminal clusters. In late summer, the fleshy berries turn deep purple and create quite a show.

GARDEN USES

This plant is not for the timid or faint of heart. Its size dictates that spikenard be used for accent or as a specimen. Use it in place of shrubs to screen an outdoor sitting area in summer. Most shade plants make suitable companions as they crouch at the feet of this gentle giant. Spikenard also makes an interesting textural background to an intricate garden planting, either alone or in combination with large ferns such as ostrich fern (*Matteuccia struthiopteris*), interrupted fern (*Osmunda claytoniana*), and male fern (*Dryopteris filix-mas*).

GROWING AND PROPAGATION

Plant spikenard in humus-rich, moist soil in partial to deep shade. Plants grow well in dense, season-long shade. Set out young plants in their permanent spot and leave plenty of room for them to spread. Established plants are very difficult to move. Propagate from root cuttings taken in spring or fall. Sow ripe seed indoors or out. Self-sown seedlings will appear. They develop slowly.

A DAINTIER *ARALIA* SPECIES

Wild sarsaparilla (*Aralia nudicaulis*) forms open colonies from fleshy, creeping rhizomes that have a pungent aroma. The leaves emerge, shiny and red-tinged, in a whorl of three pinnately compound segments atop a slender, 1- to 1½-foot-tall stalk. Spherical clusters of small, green flowers bloom in late spring or early summer. Plant in average to rich, moist soil in light to full shade. Established plants tolerate dry soil. Zones 3 to 8.

Arisaema triphyllum

Jack-in-the-Pulpit

Pronunciation	ar-iss-EE-muh try-FILL-um
Family	Araceae, Arum Family
USDA Hardiness Zones	3 to 9
Native Habitat and Range	Rich deciduous woods, bottomlands, and streamsides in a variety of soils from Nova Scotia to Manitoba, south to Florida and Louisiana

Jack-in-the-Pulpit

DESCRIPTION

Jack-in-the-pulpit is often the plant that first catches the eye of neophyte wildflower enthusiasts. Its unusual flower resembles a green calla lily, with a showy purple and green hood (spathe) surrounding a short, tonguelike central column (spadix). The spathe droops at the tip to hide the short spadix. Bright red berries form in late summer and persist until eaten or decayed. Seedlings have but one leaf; mature plants have pairs of three-lobed, 8- to 12-inch-long leaves. Plants grow 1 to 3 feet tall.

GARDEN USES

Use Jack-in-the-pulpit in the woodland garden with wildflowers such as bloodroot (*Sanguinaria canadensis*), mayapple (*Podophyllum peltatum*), spring beauty (*Claytonia virginica*), merrybells (*Uvularia* spp.), and celandine poppy (*Stylophorum diphyllum*). Combine in a shaded recess with hostas and ferns, or punctuate a low carpet of wild ginger (*Asarum* spp.) or foamflowers (*Tiarella* spp.) with Jack-in-the-pulpit's upright form.

GROWING AND PROPAGATION

Plant these woodland wildflowers in humus-rich, consistently moist soil in full to partial shade. They tolerate very deep shade and poorly drained soils. On drier sites, plants will go dormant when water is scarce. Rust may form orange pimples on the undersides of leaves in spring. In severe cases, the entire plant may be covered. Rust deforms and eventually kills infected plants. Remove and destroy affected plants as soon as the disease is noticed. Propagate Jack-in-the-pulpit by sowing cleaned seed outdoors when ripe. An easy method is to push the seeds into the ground around the parent plant. Seedlings take several years to reach flowering size.

AN *ARISAEMA* SPECIES FOR GARDEN ACCENT

Green dragon (*Arisaema dracontium*) grows 1 to 3½ feet tall. A single, horseshoe-shaped leaf, 1 to 2 feet wide with seven to nine leaflets, towers above the green spathe and long, tonguelike spadix for which the plant is named. Plant in moist to wet, acidic to neutral soil in full sun to partial shade. Zones 4 to 9.

Aruncus dioicus

Goat's Beard

Pronunciation	uh-RUN-kuss die-OH-ih-kuss
Family	Rosaceae, Rose Family
USDA Hardiness Zones	3 to 8
Native Habitat and Range	Open deciduous or mixed coniferous woods, shaded slopes, clearings, woodland edges, and open roadsides in rich soil from Pennsylvania and Iowa, south to North Carolina and Arkansas, as well as the Pacific Northwest

Goat's Beard

DESCRIPTION

Goat's beard is a showy perennial that grows to 6 feet tall with 1- to 3-foot, intricately divided leaves with many broad, quilted leaflets. The robust stems are crowned with open, fuzzy white flower plumes that resemble those of astilbes. Male and female flowers are borne on separate plants. Multiple, fuzzy stamens make the male flowers more showy, but the female flowers are also attractive, and nurseries do not distinguish between male and female plants. The frothy 2-foot plumes open in late spring and early summer. Plants grow from sturdy, many-eyed crowns with fibrous roots. 'Child of Two Worlds' is more compact, reaching 3 to 4 feet tall; 'Kneiffii' is 3 feet tall with deeply cut, ferny foliage.

GARDEN USES

In the garden, goat's beards make striking specimens or accent plants. Display them singly or in small groups to mark the entrance to a woodland trail or along a shaded path. They combine well with flowering shrubs such as viburnums and smooth hydrangea (*Hydrangea arborescens*). In open shade, combine them with ferns and wildflowers such as alumroots (*Heuchera* spp.), starry campion (*Silene stellata*), bowman's-root (*Gillenia trifoliata*), and false Solomon's seal (*Smilacina racemosa*). Ferns, grasses, and sedges (*Carex* spp.) are ideal companions. In beds and mixed borders, treat goat's beards like shrubs. Combine them with bold perennials such as garden phlox (*Phlox paniculata*), bee balms (*Monarda* spp.), queen-of-the-prairie (*Filipendula rubra*), and globe thistle (*Echinops ritro*). Where space is limited, plant a cultivar with a more compact form.

GROWING AND PROPAGATION

Plant goat's beard in moist, humus-rich soil in light to partial shade. Plants will tolerate full sun if kept evenly moist. In deep shade, plants will bloom sparsely and will be less substantial than open-grown plants. Place plants at least 4 feet apart when planting in groups. The thick roots are difficult to move once the plants are established. If division is necessary, lift clumps in the spring and use a sharp knife or shears to cut the crown into sections. Leave at least one eye (bud) per division. Sow fresh seed outdoors or inside with bottom heat. Seedlings take 2 to 3 weeks to germinate and develop quickly.

Asarum canadense

Canada Wild Ginger

Pronunciation	uh-SAH-rum can-uh-DEN-see
Family	Aristolochiaceae, Birthwort Family
USDA Hardiness Zones	3 to 9
Native Habitat and Range	Deciduous or, rarely, mixed coniferous forests, bottomlands, and slopes in rich, nearly neutral or acidic soils from New Brunswick to Ontario, south to North Carolina, Alabama, and Arkansas

DESCRIPTION

Canada wild ginger spreads quickly via shallow, creeping rhizomes to cover the ground with satiny, broadly heart-shaped deciduous leaves up to 8 inches wide. The flowers are reddish brown with three flaring, long-pointed lobes. They open just as the leaves are expanding in spring but are quickly hidden beneath the foliage.

GARDEN USES

Combine Canada wild ginger with woodland plants such as bloodroot (*Sanguinaria canadensis*), Allegheny foamflower (*Tiarella cordifolia*), baneberries (*Actaea* spp.), and ferns. Their luscious foliage is perfect for hiding the empty spots left by ephemeral species such as Virginia bluebells (*Mertensia virginica*), trout lilies (*Erythronium* spp.), false rue anemone (*Isopyrum biternatum*), and shooting-stars (*Dodecatheon* spp.). Canada wild ginger will form large colonies, which create an excellent carpet effect under trees and shrubs where root competition makes it hard for other plants to grow. The flowers are best seen when clumps are planted on a slope or at the edge of rock walls where the expanded foliage does not obscure the flowers.

GROWING AND PROPAGATION

Wild gingers are care-free garden dwellers. Plant them in consistently moist, humus-rich soil in partial to full shade.

Canada Wild Ginger

Canada wild ginger is the only *Asarum* species that tolerates limy soils. It spreads rapidly to form a dense, tidy groundcover of exceptional beauty. Divide overgrown clumps in early spring as the leaves are emerging or as they go dormant in fall. Sow fresh seed inside or outside when ripe in midsummer. Seedlings germinate quickly but plants develop slowly. They will freely self-sow in the garden.

ANOTHER WILD GINGER SPECIES

Mottled wild ginger (*Asarum shuttleworthii,* also sold as *Hexastylis shuttleworthii*) has 2- to 3-inch, glossy, heart-shaped, evergreen leaves mottled in creamy white. Plants grow from a slow-creeping rhizome to form dense clumps. The 2-inch, purple flowers have broad, rounded lobes mottled with white. 'Callaway' has smaller leaves and spreads rapidly. Deciduous and mixed coniferous woods from Virginia to Georgia and Alabama. Zones 4 to 9.

Asclepias speciosa

Showy Milkweed

Pronunciation	uh-SKLEE-pea-us spee-see-OH-suh
Family	Asclepiadaceae, Milkweed Family
USDA Hardiness Zones	2 to 9
Native Habitat and Range	Prairies, savannas, and roadsides in consistently moist soils from Manitoba to British Columbia, south to Iowa and Oklahoma

DESCRIPTION

Showy milkweed is a compact plant that grows 1 to 3 feet tall. Clusters of ½-inch, starry, rose-purple flowers are held upright in the leaf axils in early summer. The pale green, opposite leaves are 4 to 8 inches long.

Showy Milkweed

GARDEN USES

Use a generous planting of showy milkweed in a prairie or meadow with native grasses and wildflowers. The spectacular flowers make quite an impression when seen for the first time. Combine them with gayfeathers (*Liatris* spp.), wild bergamot (*Monarda fistulosa*), flowering spurge (*Euphorbia corollata*), downy phlox (*Phlox pilosa*), golden Alexanders (*Zizia* spp.), purple prairie clover (*Dalea purpurea*), and grasses. In formal gardens, plant them with yarrows, baby's-breath (*Gypsophila paniculata*), Siberian iris (*Iris sibirica*), catmints (*Nepeta* spp.), purple coneflowers (*Echinacea* spp.), and baptisias (*Baptisia* spp.).

GROWING AND PROPAGATION

Plant showy milkweed in rich, evenly moist soil in full sun or light shade. Established plants tolerate dry soils. Plants spread by underground runners and may be invasive in small gardens. Plant them in a bottomless container to control their spread or remove unwanted shoots as they appear. This species is particularly attractive to monarch butterfly caterpillars, which will feed on the foliage. As much as possible, ignore the damage and enjoy the resulting color of butterflies in your garden. Propagate from tip cuttings in May or June, or sow fresh seed indoors in a warm (70°F), sunny spot. Seedlings germinate in 3 to 5 weeks and develop quickly.

ANOTHER *ASCLEPIAS* SPECIES FOR MOIST SOILS

Swamp milkweed (*Asclepias incarnata*) carries flat clusters of pale rose to deep rose-purple flowers atop 3- to 5-foot stems. The opposite leaves are 4 to 6 inches long and lance-shaped. Stems grow in loose clumps from a creeping rhizome. Swamp milkweed favors moist to wet soils. Low meadows, prairies, wet ditches, and marshes from Nova Scotia to Saskatchewan, south to Florida and New Mexico. 'Ice Ballet' is a ghostly beautiful, creamy white selection. Zones 3 to 8.

Asclepias tuberosa

Butterfly Weed

Pronunciation	uh-SKLEE-pea-us too-ber-ROW-suh
Family	Asclepiadaceae, Milkweed Family
USDA Hardiness Zones	3 to 9
Native Habitat and Range	Dry meadows, woodland edges, sand barrens, dunes, roadsides, and prairies in a variety of soils from New England to Minnesota, south to Florida and Mexico

DESCRIPTION

Butterfly weed is always at the center of a cloud of whirling butterflies, bees, and other insects that find the flowers irresistible. Butterfly weed forms dense clumps of leafy stems to 3 feet tall from brittle, fleshy taproots. The 4- to 5-inch, lance-shaped leaves crowd hairy stems topped in summer with broad flat clusters of fiery orange, red, or, rarely, yellow flowers.

GARDEN USES

Butterfly weed adds fiery color to the early summer garden. Use a generous planting in restored grasslands or stylized prairie gardens. In formal perennial borders, position butterfly weed in the front or middle of the border in combination with yarrows, blanket flower (*Gaillardia aristata*), 'East Friesland' sage (*Salvia* × *superba* 'East Friesland'), and silver-leaved artemisias. The flowers are particularly showy combined with complementary shades of blue and purple flowers.

GROWING AND PROPAGATION

Plant butterfly weed in average, moist or dry soils in full sun or light shade. Plants can take full, scorching sun and dry soil without so much as a wilted leaf. Set out young plants in their permanent positions—once the deep taproot is established, plants are difficult to move. Spider mites may attack plants in hot, dry situations. Wash them off with a strong stream of water or spray with insecticidal soap. Propagate butterfly weed from seed or root cuttings as described under *Asclepias speciosa*. Whether seeds are sown indoors or out, transplant seedlings of butterfly weed as soon as the second set of true leaves is formed. Delayed transplanting may damage the taproot.

'Gay Butterflies' is a seed-grown strain of mixed yellow-, red-, and orange-flowering plants. Most plants deliver good, deep colors. If you choose plants in flower, you can be sure of the specific color, but otherwise you will get a mix.

Butterfly Weed

OTHER *ASCLEPIAS* SPECIES FOR PRAIRIES AND MEADOWS

Green milkweed (*Asclepias viridiflora*) has sparse, 5-inch, lance-shaped leaves on 2- to 3-foot-tall stems borne singly or in open clumps from a deep taproot. Small green flowers with reflexed petals and protruding horns bloom in tight clusters in early to midsummer. Grow in average or sandy dry soil in full sun. Dry prairies, savannas, barrens, meadows, and clearings from New England to Manitoba, south to Georgia and New Mexico. Zones 3 to 9.

Whorled milkweed (*Asclepias verticillata*) is a slender, airy plant with tight whorls of 2- to 4-inch needlelike leaves. Small, creamy white flowers are held in tight, spherical clusters near the top of the stem in mid- to late summer. Plants spread by rhizomes to form large colonies. Grow in average to rich, well-drained soil in full sun or light shade. Zones 3 to 9.

Aster divaricatus

White Wood Aster

Pronunciation	AS-ter dih-var-ih-KAH-tus
Family	Asteraceae, Aster Family
USDA Hardiness Zones	4 to 8
Native Habitat and Range	Deciduous woods, woodland coves, clearings, and roadsides from New England to Ohio, south to Georgia and Alabama

White Wood Aster

DESCRIPTION

White wood aster is valued for its mounds of late summer bloom in shady sites. The starry, ¾-inch, daisylike white flowers form broad clusters atop wiry black stems that grow 1 to 1½ feet tall with 7-inch, broadly arrow-shaped leaves with coarse, jagged teeth. Plants spread by slow-creeping runners from fibrous-rooted crowns.

GARDEN USES

White wood aster brightens the shaded recesses of the late summer woodland. The foliage is attractive all season as the bloom stalks elongate. Plant them with species that go dormant after flowering such as Virginia bluebells (*Mertensia virginica*), mayapple (*Podophyllum peltatum*), and wild leek (*Allium tricoccum*). For late summer interest, combine them with goldenrods and ferns. In beds and borders, plant them in shade with hostas, lungworts (*Pulmonaria* spp.), epimediums (*Epimedium* spp.), and sedges (*Carex* spp.) or in sunny spots with other asters, grasses, phlox, and mums.

GROWING AND PROPAGATION

White wood aster and other woodland aster species tolerate deep shade but bloom best in light to partial shade. They prefer moist humus-rich soil but tolerate dry soil. Plants spread to form attractive groundcovers. Many aster species are susceptible to aster wilt, a fungus that attacks the roots. Good drainage is the best preventive. Dig up and destroy severely infected plants. Divide overgrown clumps as necessary in early spring or after flowering. Take stem cuttings in early summer. All asters are easily grown from seed, but germination can be uneven. Sow ripe seed outdoors or indoors in flats. Place watered flats in the refrigerator for 4 to 6 weeks and then return them to a sunny position.

OTHER WOODLAND ASTERS

Bigleaf aster (*Aster macrophyllus*) creates a lush groundcover of 8-inch, heart-shaped leaves. The tall, leafy bloom stalk is crowned by a broad, mounded cluster of 1-inch white, pale blue, or lavender flowers. The foliage turns rose-pink to purple in autumn. Grows best in moist, humus-rich soil in sun or partial shade but tolerates dry sites and deep shade. Deciduous or coniferous woodlands, clearings, and woodland edges from Quebec to Minnesota, south to Georgia and Indiana. Zones 3 to 7.

Heart-leafed aster (*Aster cordifolius*) has branched, upright stems 3 to 5 feet tall that bear 5-inch, lance-shaped, toothed leaves with broad, heart-shaped bases. In late summer and early autumn the plant erupts into a dome of white to sky blue, ¾-inch flowers. Plant in moist, humus-rich soil in light to partial shade. Open woods and clearings from Nova Scotia to Minnesota, south to Georgia and Missouri. Zones 4 to 8.

Aster laevis

Smooth Aster

Pronunciation	AS-ter LEE-vis
Family	Asteraceae, Aster Family
USDA Hardiness Zones	2 to 7
Native Habitat and Range	Dry to moist meadows, prairies, woodland edges, and roadsides from Maine to British Columbia, south to Georgia and New Mexico

Smooth Aster

DESCRIPTION

Smooth aster is a showy, floriferous species with pale lavender-blue flowers in open clusters on 2- to 3½-foot plants. This is one of the last asters to bloom in autumn. The 3- to 5-inch, smooth, blue-green basal leaves are narrowly heart-shaped with clasping leaf bases. The stems bear broadly lance-shaped leaves that become smaller as they ascend the stem. Stems and leaves are often tinged with purple. 'Blue Bird' is a compact selection with deep sky blue flowers.

GARDEN USES

The late-season flowers of smooth aster add a touch of blue to the golden and russet colors of late autumn. Combine them with toad lilies (*Tricyrtis* spp.), monkshoods (*Aconitum* spp.), ornamental grasses, and the colorful foliage and berries of shrubs such as beautyberries (*Callicarpa* spp.), viburnums, hollies, fothergillas (*Fothergilla* spp.), and sweetspire (*Itea virginica*). Plants are lovely in drifts in prairie and meadow gardens.

GROWING AND PROPAGATION

Plant smooth aster in average to rich, moist but well-drained soil in full sun or light shade. Established plants are drought-tolerant. Tall stems may flop or recline in shade. Plant them among grasses, shrubs, and other plants to help prop them up. Divide in spring and take cuttings in early to midsummer.

OTHER *ASTER* SPECIES FOR PRAIRIES AND MEADOWS

Azure aster (*Aster oolentangiensis*, formerly *Aster azureus*) is similar to smooth aster, but the flowers are smaller and the inflorescence is broader and more open. The basal leaves are narrowly heart-shaped and rough like sandpaper. The stem leaves are much smaller. Plants grow 2 to 5 feet tall and bloom in mid- to late autumn. Grows best in average, dry to moist, well-drained soil in full sun. Found in dry or moist prairies, meadows, rocky slopes, and open woods and on roadsides from Ontario to Saskatchewan, south to New York and Texas. Zones 3 to 9.

Silky aster (*Aster sericeus*) has wiry, brittle stems sparsely clothed in 1½-inch, oblong leaves coated with silky, silvery hairs. The lower leaves drop as the stems elongate. The upper branches have smaller, tightly packed leaves that set off the 1½-inch, deep violet flowers in late autumn. Plant in average to sandy, well-drained soil in full sun. In rich or moist soils, plants have a more open habit and tend to flop. Plants grow best in lean, well-drained soils. Combine with delicate prairie plants such as dotted blazing star (*Liatris punctata*) and little bluestem (*Schizachyrium scoparium*). Found on gravel prairies, slopes, open woods, and savannas from Ontario to Minnesota, south to Alabama and Texas. Zones 3 to 9.

Aster lateriflorus

Calico Aster

Pronunciation	AS-ter lat-er-ih-FLOOR-us
Family	Asteraceae, Aster Family
USDA Hardiness Zones	4 to 9
Native Habitat and Range	Meadows, open woods, roadsides, dunes, and waste places, often in poor or sandy soils from New Brunswick to Minnesota, south to Florida and Texas

Calico Aster

DESCRIPTION

Calico aster is a bushy plant that bears hundreds of small white flowers on branching, leafy stems. The autumn flowers cluster along the horizontal branches of the broad inflorescence, creating an attractive presentation. Plants grow 2 to 4 feet tall with narrow, lance-shaped leaves 4 to 6 inches long. 'Horizontalis' grows to 2½ feet with burgundy-red foliage and white flowers with reddish centers. 'Prince' is a 3-foot English selection with purple-red leaves and stems and white flowers with raspberry-red, conical centers. Plant in full sun for the best color.

GARDEN USES

The frothy white sprays of calico aster add a distinctive note to the autumn garden. Plant them in groups for waves of flowers in a meadow with goldenrods and grasses. In formal situations, choose one of the compact cultivars for the front or middle of the border. Contrast the small flowers with the larger blooms of mums, New England aster (*Aster novae-angliae*), turtleheads (*Chelone* spp.), toad lilies (*Tricyrtis* spp.), and sunflowers. In England I saw them used in place of a low hedge to outline the front edge of a formal garden bed filled with taller perennials.

GROWING AND PROPAGATION

Plant calico aster in average to rich, well-drained soil in full sun or light shade. In dry sites, the lower leaves may be shed as the stems elongate. Plants form broad, dense clumps in just a few years. They spread by slow-creeping runners but are never invasive. Divide oversize clumps in early spring or after flowering. Take cuttings in early summer. Easy to grow from seed.

OTHER SMALL-FLOWERED WHITE *ASTER* SPECIES

Heath aster (*Aster ericoides*) is a densely flowered aster with hundreds of white or pale blue flowers borne on stiff branches. The 1- to 3-foot plants are more upright than calico aster. The narrow, 1- to 3-inch leaves are needlelike, especially near the tops of the stems. 'Blue Star' has sky blue flowers. 'Esther' has pale pink flowers on 2- to 3-foot plants. 'Cinderella' has white flowers with reddish centers. Plant in average to rich, moist soil in full sun. Found from Maine to Manitoba, south to Pennsylvania and New Mexico. Zones 3 to 8.

Upland white aster (*Aster ptarmicoides*) is a charming plant with dense rosettes of attractive 8-inch, deep green leaves with narrow, flat blades. The flat-topped, branched inflorescence sports showy, 1-inch, pure white flowers with bright yellow centers that open in late summer. Plants form tight, tufted clumps. They are perfect for rock gardens and containers. Plant in average to rich, well-drained soil in full sun or light shade. Found in open woods and on prairies and rocky slopes, usually in limy soils from Quebec to Saskatchewan, south to Georgia, Arkansas, and Wyoming.

Aster novae-angliae

New England Aster, Michaelmas Daisy

Pronunciation	AS-ter NO-vay ANG-lee-eye
Family	Asteraceae, Aster Family
USDA Hardiness Zones	3 to 8
Native Habitat and Range	Low meadows, wet prairies, streamsides, pond margins, and the edges of wetlands from New England and North Dakota, south to Alabama and New Mexico

New England Aster

DESCRIPTION

New England aster is a tall, stately plant with 5-inch, blunt-tipped, lance-shaped leaves that clasp the hairy, 3- to 6-foot-tall stems. Mature plants produce dense clumps from woody, fibrous-rooted crowns. The 1½- to 2-inch, lavender to purple, late-summer flowers have bright yellow centers. Flowers may vary in color from white to pink and rose.

GARDEN USES

Mix New England asters with fall-blooming perennials such as bush clover (*Lespedeza* spp.), Japanese anemones (*Anemone* × *hybrida*), common sneezeweed (*Helenium autumnale*), butterfly bushes (*Buddleia* spp.), and ornamental grasses. Most selections are suited to the back of the border, but a few dwarf cultivars work well in the front. In meadows and prairie gardens, combine them with turtleheads (*Chelone* spp.), goldenrods, sunflowers, Joe-Pye weeds (*Eupatorium* spp.), rosinweeds (*Silphium* spp.), and other asters.

GROWING AND PROPAGATION

Plant New England asters in moist, humus-rich soil in full sun or light shade. Plants tolerate consistently moist or wet soil. On dry sites, plants lose their lower leaves by midsummer. To keep plants compact and erect, shear them back to 8 to 12 inches in June. They will resprout into bushy clumps. Plants become quite large with age. Divide every 3 to 4 years in spring. Plants may need staking. New England aster is susceptible to powdery mildew, which turns the leaves dull gray. Spray with sulfur fungicide every 10 days during warm, wet weather. Thin crowded stems to promote air circulation, and keep water off the leaves. Destroy infected leaves. Propagate from seed or by 4- to 6-inch tip cuttings taken in late spring or early summer. Sow ripe seed outdoors or indoors in flats. Place watered flats in the refrigerator for 4 to 6 weeks and then return them to a sunny position. Seedlings germinate and develop quickly.

RECOMMENDED CULTIVARS OF NEW ENGLAND ASTER

'Barr's Pink' blooms late with deep rose-pink flowers on tall, 3- to 4-foot stems.

'Hella Lacy' has royal purple flowers on tight, 3- to 4-foot clumps.

'Purple Dome' is a new dwarf selection with royal purple flowers on 2-foot, late-flowering clumps.

'September Ruby' has deep ruby-red flowers on floppy, 3- to 5-foot plants.

Aster novi-belgii

New York Aster, Michaelmas Daisy

Pronunciation	AS-ter NO-vee BEL-jee-eye
Family	Asteraceae, Aster Family
USDA Hardiness Zones	3 to 8
Native Habitat and Range	Low woodlands, ditches, and wetland borders along the coast from Newfoundland to Georgia

New York Aster

DESCRIPTION

New York aster is a variable, fall-blooming species with 4- to 7-inch smooth, lance-shaped leaves and 1- to 1½-inch flowers with wide, densely packed, blue to white ray petals and yellow centers. The wild form of the species grows to 6 feet, but its hybrids are quite variable, ranging from 1 to 4 feet.

GARDEN USES

New York aster's abundant fall flowers combine well in the border with goldenrods, late-flowering astilbes such as *Astilbe chinensis*, and ornamental grasses. The mounding foliage of cranesbills (*Geranium* spp.) makes an attractive cover for asters' bare lower stems. Cultivars vary in height; many are selections with *A. dumosus*. Compact cultivars form flower-covered mounds for the front of the border, while taller selections provide a floral backdrop for lower-growing plants. In meadows and prairie gardens, combine the taller wild form of New York aster with turtleheads (*Chelone* spp.), ironweeds (*Vernonia* spp.), common sneezeweed (*Helenium autumnale*), Joe-Pye weeds (*Eupatorium* spp.), and hibiscus.

GROWING AND PROPAGATION

Plant New York asters in evenly moist, humus-rich soil in full sun or light shade. Plants grown in partial shade have a more open habit, and taller selections may flop. New York asters tolerate seaside conditions and saline soils. Moisture-stressed plants may drop their lower leaves during the summer and produce fewer flowers. Divide in spring every 3 to 4 years to reduce clumps. New York aster is susceptible to powdery mildew, particularly when plants are crowded or stressed. Leaf spot disease may also cause them to shed their lower leaves. Spray with sulfur fungicide every 10 days during warm, wet weather. Thin crowded stems to promote air circulation and keep water off the leaves. Destroy infected leaves. Propagate by seed or tip cuttings, as described under *Aster novae-angliae*.

RECOMMENDED CULTIVARS OF NEW YORK ASTER

'Ada Ballard' has double, lavender-blue flowers on 3-foot stems.

'Audrey' has lilac flowers on 1-foot plants.

'Bonningdale White' has semidouble, white flowers on 4-foot plants.

'Eventide' has 2-inch, semidouble, lavender-blue flowers on 3- to 4-foot stems.

'Jenny' has red flowers on 1-foot stems.

'Marie Ballard' is an old cultivar with double, powder blue flowers on 4-foot stems.

'Professor Kippenburg' has semidouble, lavender-blue flowers on compact, 1-foot plants.

Astilbe biternata

False Goat's Beard

Pronunciation	uh-STILL-bee bye-ter-NAH-tuh
Family	Saxifragaceae, Saxifrage Family
USDA Hardiness Zones	4 to 8
Native Habitat and Range	Rich deciduous woodlands, woodland coves, and shaded roadsides from Virginia, south to Kentucky and Georgia, in the mountains

False Goat's Beard

DESCRIPTION

False goat's beard is a handsome 3- to 5-foot woodland plant with large, 1- to 2-foot, fernlike divided foliage. The oval glossy leaflets have heart-shaped bases and toothed margins. Open, pyramidal clusters of small, creamy flowers are carried above the foliage in early summer. The tips of the flower clusters droop decoratively. The flowers fade to lime green with age and the seedheads turn rich brown.

GARDEN USES

False goat's beard is often overlooked in favor of showier *Astilbe* species, but the drooping, conical heads are attractive in the wild garden. Use them in formal and informal situations with wildflowers and traditional perennials. Combine them with ferns, irises, Solomon's seals (*Polygonatum* spp.), and shrubs at the edge of a woodland. Plant them along a stream or at the edge of a pond where their plumes are reflected in the water.

GROWING AND PROPAGATION

Like all astilbes, false goat's beard requires a moist site. Plant in consistently moist, slightly acid, humus-rich soil in light to deep shade. Dry soil and hot sun are sure death, especially in warmer zones. Provide shade from hot afternoon sun to prolong the attractive foliage display after flowering. In warmer regions, light to partial shade is mandatory. Plants do not tolerate high summer night temperatures. To keep clumps vigorous, divide them every 3 to 4 years in spring or fall and replant into well-prepared soil. Removing the flowers does not promote continued flowering, and the seedheads are attractive in their own right. Leave them standing for late-season interest. Propagate cultivars by division as seeds do not come true to their parents. Seeds are short-lived and difficult to germinate. Sow them immediately when ripe. Give them a warm (70°F), moist treatment for 2 weeks, followed by a cool (40°F), moist treatment for 4 weeks.

A PERFECT GARDEN COMPANION

Goldenseal (*Hydrastis canadensis*) has a single flower that consists only of fuzzy white stamens. The foliage is the prize of this odd plant. Use it under false goat's beard's tall stems along with spring beauties (*Claytonia* spp.), hepaticas, and irises. Zones 4 to 8.

Baptisia alba (includes *B. lactea*, *B. leucantha*, and *B. pendula*)

White Wild Indigo

Pronunciation	bap-TEEZ-ee-uh AL-buh
Family	Fabaceae, Pea Family
USDA Hardiness Zones	4 to 8
Native Habitat and Range	Open woods, meadows, prairies, and savannas in moist or dry soils from Ontario to Minnesota, south to Florida and Texas

White Wild Indigo

DESCRIPTION

White wild indigo is a species with many regional variations that are sometimes offered as separate species. In summer, plants bear long, erect spikes of white flowers like those of peas. The compound leaves have three, rounded, gray-green to bluish leaflets. Flowers give way to showy gray or brown seedheads that resemble thickened pea pods. *Baptisia alba,* or *B. lactea* (LACK-tee-uh), is a compact, 2- to 4-foot plant with leaves 1 to 2 inches long. Its white flowers are often veined in purple. Plants listed as *B. leucantha* (lew-CAN-thuh) grow 3 to 5 feet tall, with 2- to 3-inch leaves. These tall plants have pointed spikes of white flowers held well above the horizontal clump of foliage. Plants identified as *B. pendula* (PEN-djew-luh) grow 3 to 4 feet tall, with leaves 2 to 3 inches long and nodding seedpods.

GARDEN USES

White wild indigo are spectacular border plants. Place them in the middle or toward the back of the border as exclamation points among the round and mounded forms of yarrows, artemisias, asters, and phlox. Plant low, full plants such as cranesbills (*Geranium* spp.) around the base of a clump of white wild indigo to hide the naked "ankles" of the tall stalks. In prairies and meadows, plant them with their wild companions, including purple coneflowers (*Echinacea* spp.), mountain mints (*Pychnanthemum* spp.), downy phlox (*Phlox pilosa*), butterfly weed (*Asclepias tuberosa*), bowman's-root (*Gillenia trifoliata*), and prairie clovers (*Dalea* spp.). The seedpods persist throughout the fall and are quite showy in the garden or when cut and used in dried arrangements. Cut them after they color and hang them upside down to dry.

GROWING AND PROPAGATION

Wild indigo are stalwart, long-lived, easy-care perennials. Plant them in rich, moist but well-drained soil in full sun or light shade. All species are drought-tolerant once established. Wild indigo grow slowly at first but eventually spread to form huge clumps. Established plants have massive, tough roots. Set out plants when young or you may damage the taproots and stunt their growth for several years. Space individual plants at least 3 feet apart. The seeds have extremely hard seedcoats. Soak them overnight in hot water before sowing. Germination will occur over a 2- to 4-week period. Keep seedlings on the dry side to prevent damping-off, a fungal disease that attacks seedlings at the soil line, causing them to topple over.

Baptisia australis

Blue False Indigo

Pronunciation	bap-TEEZ-ee-uh aw-STRAH-lis
Family	Fabaceae, Pea Family
USDA Hardiness Zones	3 to 9
Native Habitat and Range	Open woods, moist meadows, and prairies in a variety of soils from Vermont and Pennsylvania to Indiana, south to North Carolina and Tennessee

Blue False Indigo

DESCRIPTION

Blue false indigo's pealike, deep blue flowers bloom on open spikes up to 1 foot long above the soft, blue-green foliage in late spring. Each clump produces many 2- to 4-foot stems from a tough, gnarled, deep taproot. The elongated seedpods turn charcoal gray when they ripen.

GARDEN USES

Blue false indigo never fails to attract attention, whether in a formal border or along a roadside. Use it in a bold sweep or in scattered clumps to add a splash of color to meadow and prairie plantings. In beds and borders, plant it in the company of bold flowers such as peonies, oriental poppies (*Papaver orientale*), and irises. Plant airy plants such as columbines (*Aquilegia* spp.) and bleeding hearts (*Dicentra* spp.) around its base. After flowering, the shrubby plants make an excellent background for late-blooming perennials. Contrast the blue-green foliage with fine-textured ornamental grasses. The showy seedpods add interest in the late-summer and fall garden. Chickadees love to eat the seeds in winter.

GROWING AND PROPAGATION

Blue false indigo is a durable, low-care perennial. Plants thrive in average to rich, moist but well-drained soil in full sun or light shade. Established plants are extremely drought-tolerant, looking flawless through severe drought and scorching heat. Plants will bloom well in light shade but may need staking. Place rounded peony hoops over the clumps as they emerge in early spring. Avoid moving established plants, as they grow slowly and resent disturbance. Divide in fall if necessary. You will need a sharp knife or shears to cut through the tough clumps. Leave at least one bud or "eye" per division. Propagate from seed as described under *Baptisia alba*.

ANOTHER RECOMMENDED *BAPTISIA* SPECIES

Blue false indigo (*Baptisia minor*) is a petite version of *Baptisia australis* and is often listed as a variety (*B. australis* var. *minor*). Plants grow 1 to 3 feet tall. The leaves are more dainty and the spikes of blue flowers grow to 6 inches tall. Plant in average to rich, moist soil in full sun or light shade. Found in open woods, prairies, savannas, glades, and clearings from Virginia to Nebraska, south to Georgia and Oklahoma. Zones 4 to 9.

Boltonia asteroides

Boltonia

Pronunciation	bowl-TOE-nee-uh as-ter-OY-deez
Family	Asteraceae, Aster Family
USDA Hardiness Zones	3 to 8
Native Habitat and Range	Low, open woods, wetland margins, and wet ditches from New Jersey and North Dakota, south to Florida and Texas

Boltonia

DESCRIPTION

Boltonia blooms in late summer and fall with profuse, open clusters of 1-inch white daisies with bright yellow centers carried atop 4- to 6-foot stems. The 3- to 5-inch, gray-green, willowlike foliage is attractive and neat all summer. Cultivars of boltonia are best-known in the garden; wild plants have a more open form and slightly smaller flowers. 'Pink Beauty' sports soft pink flowers in open clusters. Flower color is brighter where summers are cool. 'Snowbank' is a compact selection to 4 feet and is smothered with bright white flowers throughout the fall.

GARDEN USES

Combine boltonias with fall-blooming plants such as asters, goldenrods, Joe-Pye weeds (*Eupatorium* spp.), turtleheads (*Chelone* spp.), and grasses. In formal garden situations, plant them toward the rear of the border with Japanese anemones (*Anemone* × *hybrida*), plume poppy (*Macleaya cordata*), New England aster (*Aster novae-angliae*), sneezeweeds (*Helenium* spp.), and shrubs.

GROWING AND PROPAGATION

Plant boltonia in moist, humus-rich soil in full sun or light shade. On drier soil, plants will be smaller. Divide overgrown clumps in spring. Plants form sturdy, dense stems that seldom need staking. To keep plants short, cut them back to 10 to 12 inches in early June to encourage compact growth. Cultivars do not come true from seed. Propagate by tip cuttings taken in early summer.

A PERFECT GARDEN COMPANION

Fireweed (*Epilobium angustifolium*) is a flamboyant plant that sports flashy conical clusters of ¾-inch, rose pink to magenta flowers on 4- to 6-foot stems. The deep green leaves are lance-shaped and 3 to 6 inches long. Plants bloom all summer and form extensive colonies from creeping stems. Use them at the edge of a woodland, in a low meadow, or at pondside with boltonia and other late-blooming plants such as goldenrods and asters. Plants tolerate moist or dry soils but need cool night temperatures for best growth. Self-sown seedlings are plentiful on bare soil. 'Album' is a choice selection with pure white flowers. Open deciduous or coniferous woods, meadows, roadsides, and waste places; often very common following a fire. Found from the tundra, south in North America to New Jersey, New Mexico, and California. Zones 1 to 8.

Callirhoe involucrata

Poppy Mallow

Pronunciation	kal-lih-ROW-ee in-voh-lou-KRATE-uh
Family	Malvaceae, Mallow Family
USDA Hardiness Zones	4 to 9
Native Habitat and Range	Dry, often sandy plains, prairies, and open woods from North Dakota and Montana, south to Missouri and New Mexico; naturalized farther east

Poppy Mallow

DESCRIPTION

Poppy mallow is a sprawling to creeping plant with 1- to 1½-foot stems that grow from thick, branched taproots. The attractive, 3- to 3½-inch leaves have five to seven well-defined, toothed lobes. The deep wine-red flowers are 2½ inches across. Each blossom has five petals and resembles a teacup. Flowers bloom singly above the foliage. Plants begin blooming in mid- to late spring and flower for several months on new growth.

GARDEN USES

Use poppy mallow to knit plantings together. The trailing stems creep between or over clumps of other plants, and flowers pop up here and there. Poppy mallow forms dense clumps in rich soils, so it's best to put it at the edge of a bed or along a path. In formal gardens, combine poppy mallow with sedums, lamb's-ears (*Stachys byzantina*), ornamental onions (*Allium* spp.), yarrows, and asters. In prairie gardens, plant it with milkweeds (*Asclepias* spp.), asters, blanket flowers (*Gaillardia* spp.), spiderworts (*Tradescantia* spp.), and ornamental grasses. Poppy mallow also makes a long-flowering carpet in rock walls and rock gardens.

GROWING AND PROPAGATION

Plant in average, well-drained, loamy or sandy soil in full sun or light shade. Don't try to transplant established plants—they are difficult to move because of their taproots. Seeds need cold, moist stratification to germinate. Sow in winter, and place covered flats in the refrigerator for 4 to 6 weeks. Plants will be ready to plant outdoors by fall or the following spring.

A RELATED *CALLIRHOE* SPECIES

Winecups (*Callirhoe digitata*) is an upright to sprawling plant with 1- to 4-foot stems. The decorative leaves have five to seven narrow, fingerlike lobes, and the leaves cover the stems sparsely. The 1- to 2-inch flowers are carried singly or in clusters of a few blossoms. Flower color varies from white to light rose and wine-red. Plant in well-drained, loamy or sandy soil in full sun or light shade. Plants will grow up through large perennials or shrubs for support. Dry prairies and open woods from Missouri and Kansas, south to Texas; naturalized farther east. Zones 4 to 9.

Caltha palustris

Marsh Marigold, Cowslip

Pronunciation	KAL-thuh pal-US-tris
Family	Ranunculaceae, Buttercup Family
USDA Hardiness Zones	2 to 8
Native Habitat and Range	Wet meadows, streamsides, areas near springs, pond margins, and shallow wetlands from northern boreal areas, south to North Carolina, Indiana, and Alaska

Marsh Marigold

DESCRIPTION

Marsh marigold produces mounds of rounded or slightly heart-shaped leaves with wavy, toothed margins. The 1½-inch butter yellow flowers have five, shiny, petallike sepals and bloom in open clusters. The first flowers open as the plants emerge in early spring. The bloom stalks eventually reach 1 to 2 feet tall; the clumps have an equal spread. Marsh marigold grows from thick crowns with fleshy roots; the plants go dormant by midsummer. 'Alba' is a robust selection with white flowers; plants bloom for 3 to 4 weeks. 'Flore Pleno' (also sold as 'Multiplex') is a robust, fast-growing plant with fully double flowers that last for a week or more.

GARDEN USES

Marsh marigold is perfect for water gardens or along the low banks of streams. Combine it with wild calla (*Calla palustris*), pickerel weed (*Pontederia cordata*), and arrowhead (*Sagittaria latifolia*). In a bog garden, plant it with primroses, rodgersias (*Rodgersia* spp.), irises, and ferns. The double form is particularly lovely perched over water.

GROWING AND PROPAGATION

Plant marsh marigold in consistently moist to wet, humus-rich or loamy soil in full sun or partial shade. To grow marsh marigold in containers, plant it in rich potting mix, and cover the mix with 2 inches of pea gravel before submerging in water. Plants will grow when covered with 1 to 4 inches of water. Once flowering is complete, water is less critical. Plants go dormant about a month after flowering. This is the time to divide overgrown plants. Sow fresh seeds outdoors immediately upon ripening. Do not allow the seeds to dry out or they won't germinate easily. In any case, plants will not germinate until the following spring.

A PERFECT GARDEN COMPANION

Wild calla (*Calla palustris*) has glossy, 6-inch, broadly oval to heart-shaped leaves on thick stalks. The fleshy white, petallike spathe surrounds a knobby, conical spadix that bears the actual flowers. The flowers unfurl in late spring or early summer, just as marsh marigold's flowers are fading. Use them to extend the early summer bloom in wet soil. Plant them at the edge of a stream or pond where the colonies can freely roam. Zones 2 to 7.

Campanula rotundifolia

Bluebell, Harebell

Pronunciation	kam-PAN-yew-luh row-ton-dih-FOE-lee-uh
Family	Campanulaceae, Bellflower Family
USDA Hardiness Zones	2 to 7
Native Habitat and Range	Rocky cliffs, outcroppings, dunes, dry prairies, and savannas from northern boreal areas, south to New Jersey and Iowa and in the western mountains to Mexico

Bluebell

DESCRIPTION

Bluebells are most familiar as dainty, 2- to 3-inch plants that grow wild on cliff faces along the New England coast. The 1-inch, oval to heart-shaped basal leaves have wavy, toothed edges, and the stem leaves are lance-shaped. In a protected site with good soil, plants may grow as tall as 18 inches. The deep sky blue, 1-inch nodding flowers are bell-shaped and have five pointed lobes. Plants bloom throughout the summer.

GARDEN USES

Bluebells are so delicate that it's easy to lose them in traditional gardens. Use them in unmortared rock walls, rock gardens, and between pavers on a planted terrace. Suitable companions include thymes, pussy-toes (*Antennaria* spp.), ornamental onions (*Allium* spp.), columbines (*Aquilegia* spp.), and penstemons (*Penstemon* spp.). In dry prairie and savanna gardens, plant bluebells with dotted blazing-star (*Liatris punctata*), purple prairie clover (*Dalea purpurea*), upland white aster (*Aster ptarmicoides*), blanket flower (*Gaillardia aristata*), butterfly weed (*Asclepias tuberosa*), and prairie smoke (*Geum triflorum*).

GROWING AND PROPAGATION

Plant in average to poor, well-drained, loamy or sandy soil in full sun or light shade. Plants will grow in rich, well-drained sites but may be overpowered by more vigorous plants. You may need to divide plants in spring to keep the underground stems from spreading too much. Store seed in a refrigerator until ready to sow. Sow in early spring for summer transplants. Self-sown seedlings will appear.

ANOTHER *CAMPANULA* SPECIES

Tall bellflower (*Campanula americana*) is an annual with tall, slender stems reaching 4 to 6 feet. It is sometimes grouped with the genus *Campanulastrum*. The 6-inch, toothed leaves are oval to broadly lance-shaped. The upper third of the stem bears clusters of 1-inch, flat, starry, deep blue flowers in the leaf axils. Plants bloom for 1 to 2 months in summer. Plant in rich, moist soil in light to full shade. Once established, the plants will self-sow each year. Found in rich woods, woodland coves, bottomlands, and streamsides and on roadsides from Ontario to Minnesota, south to Florida and Oklahoma. Zones 3 to 9.

Caulophyllum thalictroides

Blue Cohosh

Pronunciation	call-oh-FILL-um thuh-lick-TROY-deez
Family	Berberidaceae, Barberry Family
USDA Hardiness Zones	3 to 9
Native Habitat and Range	Rich, moist woods, woodland coves, bottomlands, and rocky forest slopes, often in limy soils, from New Brunswick and Manitoba, south to South Carolina and Alabama

Blue Cohosh

DESCRIPTION

Blue cohosh stems have a rich purple tint when they first emerge in spring. The plants have sea green leaves composed of nine leaflets arranged in three groups of three. The flowers appear in a tight cluster at the top of the stem. Each flower has six ¼-inch, petallike sepals in shades of chartreuse and bronze. The midnight-blue berries appear in the fall. Berries are ½ inch in diameter and have a waxy sheen.

GARDEN USES

Accent the purple stalks of blue cohosh with the blooms of early spring ephemerals. Try Dutchman's breeches (*Dicentra cucullaria*), spring beauties (*Claytonia* spp.), false rue anemone (*Isopyrum biternatum*), and trout lilies (*Erythronium* spp.), all of which will go dormant after flowering. The blue cohosh will then spread its leaves horizontally and fill the area. Blue cohosh foliage is lovely in summer with ferns and persistent wildflowers. In fall, the bright yellow of goldenrods is a perfect foil for the blue berries.

GROWING AND PROPAGATION

Plant in evenly moist but well-drained, humus-rich soil in partial to full shade. Plants form multistemmed clumps in time. To divide them, lift the clumps in early fall and pull or cut the sections of the crown apart. Leave at least one eye per division. The seeds have a complex dormancy that is hard to overcome with indoor sowing. Remove the pulp and sow seeds outdoors in the fall. Plants will germinate in 1 to 2 years.

PERFECT GARDEN COMPANIONS

Miterwort or **bishop's cap** (*Mitella diphylla*) gets its common names from the resemblance of its tiny flowers and seed capsules to a pointed bishop's miter. Each flower has four branched arms that look like a lacy snowflake when magnified. Miterwort is a dainty plant with evergreen, toothed triangular leaves. The 8- to 12-inch bloom stalks have pairs of stalkless leaves just below the string of creamy white spring flowers. Choose miterwort for underplanting with blue cohosh. The plants often grow together in the woods and are perfect garden companions. Plant in humus-rich, moist but well-drained, neutral or slightly acid soil in light to full shade. Zones 3 to 8.

Wild leek (*Allium tricoccum*) has broad, sea green, straplike leaves that grow from a succulent, edible bulb. The leaves emerge in early spring and persist through the early summer. In late June or July, the 6- to 10-inch flowerstalk with green flowers magically appears. Wild leeks produce a lush leaf display that complements the soft green leaves of blue cohosh. Wild leeks thrive in average to rich, moist but well-drained soils in light sun to shade. Found with blue cohosh in rich deciduous bottomlands and forests, often in limy soils. Zones 3 to 8.

Chelone lyonii

Pink Turtlehead

Pronunciation	chee-LOW-nee lye-ON-ee-eye
Family	Scrophulariaceae, Figwort Family
USDA Hardiness Zones	3 to 8
Native Habitat and Range	Areas near springs, wet meadows and clearings, streamsides, and bogs in the southern Appalachian mountains of North Carolina, South Carolina, and Tennessee

DESCRIPTION

Pink turtleheads are bushy perennials with 1- to 3-foot, leafy, arching stems. The 4- to 7-inch, oval leaves are deep green and have toothed edges. Bright rose-pink flowers appear in clusters at the ends of stems and in the axils of the leaves on the upper third of the stems. Each inflated blossom has a small opening at the end and resembles a turtle with its mouth open. Plants bloom in late summer and early fall. The dried seedheads are also attractive.

GARDEN USES

Use pink turtlehead to add color accent in the late-summer and fall garden. It combines beautifully with asters, phlox, and goldenrods in both formal and informal gardens. In a perennial border, plant pink turtlehead with Joe-Pye weeds (*Eupatorium* spp.), chrysanthemums, toadlilies (*Tricyrtis* spp.), and Japanese anemone (*Anemone* × *hybrida*). In an informal garden or meadow, plant it in drifts with sneezeweed (*Helenium autumnale*), sunflowers, and grasses. Beside a pond, pair it with irises, great blue lobelia (*Lobelia siphilitica*), and ferns.

GROWING AND PROPAGATION

Plant pink turtlehead in evenly moist, humus-rich soil in full sun or partial shade. These versatile plants thrive in wet soil, and they will also tolerate dry soil once established. But constant dryness will stunt their growth and turn the leaf margins brown. They cannot withstand high heat; in warmer zones, plants in full sun must have constant moisture. Plants produce broad clumps with age. Divide the fleshy-rooted crowns in spring or late fall after flowering. Propagate by stem cuttings taken in early summer. Sow seeds outdoors when ripe. Indoors, sow them in winter and place in the refrigerator for 4 to 6 weeks. Afterward, put flats in the light at 60°F. Plants germinate in 2 to 3 weeks and can be transplanted in summer.

Pink Turtlehead

MORE *CHELONE* SPECIES

Rose turtlehead (*Chelone obliqua*) produces abundant rich pink to rose-red flowers on upright to arching stems. The 6-inch leaves are narrower than those of pink turtlehead. Plants grow 2 to 3 feet tall. 'Alba' has white flowers. Plant in moist to wet, humus-rich soil in full sun to partial shade. It will tolerate full shade but will not flower as heavily. Wet woods and marshes of the coastal plain from Maryland to Alabama; also from Indiana to Wisconsin and south to Arkansas. Zones 4 to 9.

White turtlehead (*Chelone glabra*) is a lovely, upright to slightly vase-shaped plant with white flowers blushed with violet. The dark green, 6-inch leaves are narrow and lance-shaped. Plants grow 3 to 5 feet tall. Give them evenly moist to wet, humus-rich soil in full sun or partial shade. Wet meadows and low woods from Newfoundland and Minnesota, south to Georgia and Alabama. Zones 3 to 8.

Chrysogonum virginianum

Green and Gold

Pronunciation	kreye-SOG-oh-num ver-jin-ee-AH-num
Family	Asteraceae, Aster Family
USDA Hardiness Zones	6 to 9, hardy in colder zones with protection
Native Habitat and Range	Open woods, clearings, and embankments from Pennsylvania and Ohio, south to Florida and Alabama

Green and Gold

DESCRIPTION

Green and gold is a low, clumping or trailing plant to 1 foot tall. The woolly stems have opposite, 1- to 3-inch-long, oval leaves. The starry, bright yellow flowers have five petals surrounding a small, brown disc. Plants begin blooming as the stems elongate in early spring and continue throughout the summer.

GARDEN USES

Green and gold makes a charming and attractive flowering groundcover. Use clumps at the entrance to a woodland path or along a shaded walk, or plant beneath shrubs and flowering trees. Combine it with wild columbine (*Aquilegia canadensis*), bowman's-root (*Gillenia trifoliata*), red fire pink (*Silene virginica*), baptisias (*Baptisia* spp.), gaura (*Gaura* spp.), and Indian pink (*Spigelia marilandica*).

GROWING AND PROPAGATION

Plant green and gold in average to rich, well-drained mineral soil in full sun or partial shade. Avoid soils with excessive organic matter, as they may promote disease. In hot areas, plants benefit from some shade. Green and gold is subject to powdery mildew, which forms a gray or whitish coating on the leaves. Thin or divide crowded plantings to improve air circulation; destroy infected leaves. Divide older clumps in early spring or autumn. Take tip cuttings in early summer. Refrigerate seed until ready to sow. Moist stratification increases germination but is not essential. Seedlings develop quickly.

RECOMMENDED SELECTIONS OF GREEN AND GOLD

Chrysogonum virginianum* var. *virginianum has upright stems and forms tight, discreet clumps. 'Pierre' is a compact selection, to 6 inches, with a long season of bloom. 'Mark Viette' is a popular, fast-growing cultivar noted for its rich green foliage and numerous large flowers. 'Springbrook' is an 8-inch groundcover with shiny leaves and a long bloom period.

The variety ***australe*** is more southern in distribution than virginianum. It has prostrate stems and spreads by runners to form broad patches. 'Allen Bush' is a vigorous groundcover with fuzzy green leaves and broad petals. This variety is subject to root rot, which causes clumps to die out. To prevent problems, avoid soils with high organic content and provide good drainage. 'Eco Lacquered Spider' belongs to the lesser-known species *C. stolonifera* but may be listed as a cultivar of *C. virginianum*. It has far-trailing stems and shiny leaves.

Chrysopsis mariana (also sold as *Heterotheca mariana*)

Maryland Golden Aster

Pronunciation krih-SOP-sis mar-ee-AH-nuh

Family Asteraceae, Aster Family

USDA Hardiness Zones 4 to 9

Native Habitat and Range Open pine woods, clearings, and roadsides in acid sandy or dry clay soils from New York and Ohio, south to Florida and Louisiana

Maryland Golden Aster

DESCRIPTION

Maryland golden aster forms 1- to 3-foot-tall mounds with 6- to 9-inch, deep green, lance-shaped leaves clothed in gray hairs. Largest at the base of the plant, the leaves get smaller as they ascend the stems. In late summer, cheery, bright yellow, ½-inch, asterlike flowers cover the plant, often persisting well into fall. The tawny seedheads resemble miniature dandelions.

GARDEN USES

Maryland golden asters are adaptable and showy border plants. Use the late-season flowers with ornamental grasses, chrysanthemums, and late phlox. They are well suited to dry soil gardens with yarrows, white sage (*Artemisia ludoviciana*), gayfeathers (*Liatris* spp.), goldenrods, and asters. They perform well in rock gardens, on dry sunny banks, and in other tough sites.

GROWING AND PROPAGATION

Maryland golden asters are tough, heat- and drought-tolerant plants. Plant them in average or sandy soil in full sun or light shade. Good drainage is essential. Plants tend to languish in rich or overly moist soils and are more susceptible to problems with fungal root rot. Plants grow slowly to form tight clumps. Divide overgrown clumps in spring. Propagate by cuttings taken in early summer or by seed. Sow seeds outdoors when ripe or indoors with 4 to 6 weeks of cold stratification.

OTHER RECOMMENDED *CHRYSOPSIS* SPECIES

Hairy golden aster (*Chrysopsis villosa*) shows geographic variations in its height and flower size. Wild plants are usually 12 to 18 inches tall, while some forms may grow to 5 feet. The cultivar 'Golden Sunshine' grows 4 to 5 feet tall with broadly lance-shaped foliage clothed in long, soft hairs. The 1½- to 2-inch, golden yellow flowers are held in broad, flat clusters from late summer through fall. Zones 4 to 9.

Narrowleaf silkgrass (*Chrysopsis graminifolia*, also listed as *Pityopsis graminifolia*) is an attractive fall-blooming plant with 6- to 18-inch, stiff, silvery, grasslike leaves in a loose rosette. The leaves along the 1- to 3-foot stems are smaller. The ½-inch, yellow flowers are carried in open clusters. Plant in sandy, well-drained soil in full sun or light shade. Found in open woods, savannas, sandhills, and dunes from Delaware and Ohio, south to Florida and Mexico. Zones 6 to 9.

Cimicifuga racemosa

Black Snakeroot, Black Cohosh

Pronunciation	sim-ih-siff-YOU-guh ray-sih-MOW-suh
Family	Ranunculaceae, Buttercup Family
USDA Hardiness Zones	3 to 8
Native Habitat and Range	Rich woods, clearings, and roadsides from Massachusetts and Missouri, south to Georgia and Tennessee

Black Snakeroot

DESCRIPTION

Black snakeroot blooms in early summer, producing tall spikelike racemes covered by dozens of ¼- to ½-inch, fuzzy, ill-scented, white flowers. Flowering spikes may grow to 6 feet tall, branching like candelabras above attractive, dark green, divided leaves that may reach 1 foot or more across. The compound foliage resembles that of astilbes and remains attractive from early spring through frost.

GARDEN USES

In beds and borders, combine them with summer perennials such as daylilies, true lilies, astilbes, and daisies. Contrast their strong vertical form with rounded plants such as garden phlox (*Phlox paniculata*), yarrows, and cranesbills (*Geranium* spp.), or use them in mass as an accent in a shrub planting. In the shaded garden, plant them with ferns, sedges (*Carex* spp.), bleeding hearts (*Dicentra* spp.), columbines (*Aquilegia* spp.), wild gingers (*Asarum* spp.), and alumroots (*Heuchera* spp.).

GROWING AND PROPAGATION

Black snakeroots are long-lived woodland plants. Give them moist, humus-rich soil in sun or shade. In warmer regions, consistent moisture and shade from afternoon sun will extend the life of the foliage, which scalds in hot sun. Plants are fairly drought-tolerant once established. They grow steadily to form large, multistemmed clumps that can remain in position for many years. Divide the tough crowns in the fall with a sharp knife, leaving at least one eye, or bud, per division. Sow fresh seeds outdoors. They have a complex dormancy and may not germinate for 2 years. Self-sown seedlings may be numerous.

ANOTHER *CIMICIFUGA* SPECIES

Summer cohosh or **American bugbane** (*Cimicifuga americana*) is a late-summer–blooming plant that closely resembles black snakeroot. The plant is smaller in stature, from 2 to 5 feet, and has single or sparsely branching inflorescences. Plant in rich, moist soil in light to partial shade. Found in rich mountain woods from Pennsylvania to North Carolina and Tennessee. Zones 4 to 8.

Claytonia virginica

Spring Beauty

Pronunciation	clay-TOE-nee-uh ver-JIN-ih-kuh
Family	Portulacaceae, Purslane Family
USDA Hardiness Zones	3 to 9
Native Habitat and Range	Rich woods, bottomlands, streamsides, and glades in a variety of soils from Nova Scotia and Minnesota, south to Georgia and Texas

Spring Beauty

DESCRIPTION

Spring beauty opens its first flowers very early in the spring and continues blooming throughout the season. Plants bear loose, open clusters of starry, five-petaled, pale pink flowers with deep pink anthers. Some plants bear pure white flowers, while others are pink with deep rose-pink veins. Plants produce dense, leafy clumps of 4- to 6-inch, succulent, grasslike basal leaves. The flowering stems bear pairs of leaves just below the flower clusters. Extensive colonies of self-sown spring beauties carpet the woods under trees' bare branches. Plants grow from fleshy, edible corms that taste like water chestnuts. The entire plant disappears soon after the last flowers fade.

GARDEN USES

Few groundcovers are as delicate and floriferous as spring beauty. Its only drawback is that it disappears in summer. Use it in shade and wildflower gardens with other ephemerals such as trout lilies (*Erythronium* spp.), toothworts (*Dentaria* spp.), and Virginia bluebells (*Mertensia virginica*), with a generous interplanting of persistent wildflowers such as bloodroot (*Sanguinaria canadensis*), wild gingers (*Asarum* spp.), baneberries (*Actaea* spp.), and ferns. Plant them with traditional bulbs such as snowdrops, species crocus, and glory-of-the-snow (*Chionodoxa luciliae*) under flowering shrubs and trees. Plant a few spring beauties in your lawn, then let them seed freely through the grass to form a flowering carpet in early spring. Leave the grass unmowed until they go dormant.

GROWING AND PROPAGATION

Plant in moist, rich soil in sun or shade. Take care not to dig into dormant clumps. Plants that disappear mysteriously or fail to emerge may have fallen prey to rabbits, which savor all parts of the plant. Propagate by cutting the corms into sections as the foliage is yellowing. Leave at least one eye, or bud, per section. Plants self-sow freely and are easily transplanted.

ANOTHER RECOMMENDED *CLAYTONIA* SPECIES

Carolina spring beauty (*Claytonia caroliniana*) is smaller than Virginia spring beauty and has broadly lance-shaped to oval leaves and ½-inch flowers. The clumps are more delicate and produce fewer leaves and flowering stems. Flower color is generally white to pale pink, but flowers may be rose-pink in some areas of the country. Plants spread more slowly than Virginia spring beauty but in time will form dense carpets. Plant in rich, moist soil in sun or shade. Found in woods, woodland coves, and bottomlands from Nova Scotia and Minnesota, south around the Great Lakes and in the mountains to Tennessee and North Carolina. Zones 3 to 8.

Coreopsis grandiflora

Large-Flowered Tickseed

Pronunciation	core-ee-OP-sis gran-dih-FLOOR-uh
Family	Asteraceae, Aster Family
USDA Hardiness Zones	4 to 9
Native Habitat and Range	Open woods, clearings, meadows, and roadsides in the southeastern United States from North Carolina to Mississippi; naturalized elsewhere

Large-Flowered Tickseed

DESCRIPTION

Large-flowered tickseed is a cheery old-fashioned perennial with clumps of 3- to 6-inch, lance-shaped, entire or three- to five-lobed basal and stem leaves. In summer, numerous 2½-inch, deep yellow, daisylike flowers bloom atop 1- to 2-foot stems.

GARDEN USES

Tickseeds make excellent border plants and are great for the cutting garden, too. Combine them with yarrows, salvias, balloon flower (*Platycodon grandiflorus*), and catmints (*Nepeta* spp.). In meadows and informal settings, plant them with butterfly weed (*Asclepias tuberosa*), asters, purple coneflower (*Echinacea purpurea*), goldenrods, and grasses.

GROWING AND PROPAGATION

Tickseeds are durable, easy-care perennials. Plant them in average to rich, moist soil in full sun or light shade. They are quite drought-tolerant once established. Overly rich soils promote flopping. Remove spent flowers regularly to prolong flowering. Divide overgrown plants in spring or fall. Take stem cuttings in early summer. They root quickly and can be planted out in the same season. Sow ripe seeds indoors under warm (70°F), moist conditions. Seeds germinate in 2 to 4 weeks.

RECOMMENDED CULTIVARS OF LARGE-FLOWERED TICKSEED

'Early Sunrise' is compact with double flowers.

'Goldfink' is compact (9 inches) with single flowers.

'Sunray' has 2-inch, semidouble, golden yellow flowers on 16- to 18-inch stems.

ANOTHER *COREOPSIS* SPECIES

Lance-leaved coreopsis (*Coreopsis lanceolata*) is similar to *C. grandiflora* with mostly basal, 2- to 6-inch leaves, with stem leaves reduced in size. Flowering stems are 1 to 2 feet tall. Plant in average, sandy or loamy, well-drained soil in full sun or light shade. Listed cultivars may be hybrids of *C. lanceolata* and *C. grandiflora*. Dry, open woods, clearings, lake dunes, and meadows from the Great Lakes, south to Florida and New Mexico. Zones 3 to 8.

Coreopsis verticillata

Threadleaf Coreopsis

Pronunciation	core-ee-OP-sis ver-tih-sill-AH-tuh
Family	Asteraceae, Aster Family
USDA Hardiness Zones	3 to 9
Native Habitat and Range	Dry, open woods, clearings, pine barrens, and roadsides from Maryland, south to Florida and Arkansas

Threadleaf Coreopsis

DESCRIPTION

Threadleaf coreopsis is an airy, rounded plant from 1 to 3 feet tall with leaves divided into three threadlike lobes. The 1- to 2-inch, starry flowers are butter to golden yellow. They are borne in profusion for several months in summer. In time, plants form broad, dense clumps with many stiff, leafy stems. They spread by creeping underground stems with fibrous roots. Rose coreopsis (*Coreopsis rosea*) is a closely related species with delicate, spreading stems and rose-pink flowers.

GARDEN USES

Threadleaf coreopsis is a popular perennial and a recipient of the Perennial Plant of the Year Award. Use it in formal gardens and meadows. These plants are perfect for the front of the border with cranesbills (*Geranium* spp.), yarrows, daylilies, lamb's-ears (*Stachys lanata*), and purple coneflowers (*Echinacea* spp.). Combine them with grasses or use a mass planting to front a shrub border. They are also well suited to rock gardens. The soft yellow flowers of 'Moonbeam' combine well with blue, pink, and lavender flowers.

GROWING AND PROPAGATION

Plant threadleaf coreopsis in average to rich, moist but well-drained soil in full sun or light shade. Plants are drought-tolerant once established. Old clumps will eventually die out at the center. Divide them and replant in amended soil. Take tip cuttings in early summer.

SOME CULTIVARS OF THREADLEAF COREOPSIS

'Golden Showers' grows 2 feet tall with 2-inch, deep yellow flowers.

'Moonbeam' is a spreading plant that grows from 1 to 2 feet wide with 1- to 1½-inch, pale yellow flowers from early summer through autumn. Plants are more open in form than other selections and may be of hybrid origin.

'Zagreb' is a compact, 8- to 18-inch selection similar to 'Golden Showers'.

ANOTHER FINE-TEXTURED *COREOPSIS* SPECIES

Calliopsis (*Coreopsis tinctoria*) is an annual species with 1- to 4-foot, branched stems sporting dozens of showy 1¼-inch, yellow, orange, and brick red flowers. The leaves have threadlike dissections and the plants grow from sparsely rooted crowns. Plant in dry to moist, well-drained soil in full sun. Found in open woods, prairies, waste places, and plains and along roadsides from Saskatchewan and Washington, south to Louisiana and California. Naturalized elsewhere and frequently planted in roadside projects. Zones 3 to 10.

Cornus canadensis

Bunchberry

Pronunciation	CORE-nus kan-uh-DEN-sis
Family	Cornaceae, Dogwood Family
USDA Hardiness Zones	1 to 7
Native Habitat and Range	Forested bogs, boreal and mixed coniferous forests, open woods in acid humus from northern boreal areas, south to New Jersey and Minnesota, the mountains in West Virginia, and in the Rocky Mountains to New Mexico and California

DESCRIPTION

Bunchberry is known in northern regions around the world for its bright summer flowers and merry red berries. Plants are sub-shrubs that appear as perennials. They grow from creeping, wiry stems to form extensive colonies of plants topped by 1-inch, glowing white flowers. Each flower consists of four petallike bracts surrounding a button of small, petalless flowers. The flowers sit atop a whorl of four to six 6-inch oval leaves on 4- to 9-inch stems. Plants and flowers vary in size, depending on site.

GARDEN USES

Bunchberry brightens shaded recesses in the wild garden with goldthread (*Coptis groenlandica*), wintergreen (*Gaultheria procumbens*), starflower (*Trientalis borealis*), shortia (*Shortia* spp.), Canada mayflower (*Maianthemum canadense*), clintonias (*Clintonia* spp.), and ferns. Use them as a groundcover under acid-loving shrubs such as azaleas (*Rhododendron* spp.), leucothoes (*Leucothoe* spp.), and fothergillas (*Fothergilla* spp.).

GROWING AND PROPAGATION

Plant in cool, moist, humus-rich, acid soil in partial to full shade. Plants are difficult to keep going under less than optimum conditions. They are intolerant of summer heat and drought. Propagate by division in spring. Sow cleaned seeds outdoors in autumn. Seedlings will appear the following summer but take several years to bloom.

Bunchberry Flowers

Bunchberry Berries

Dalea purpurea

Purple Prairie Clover

Pronunciation	DAY-lee-uh purr-PURR-ee-uh
Family	Fabaceae, Pea Family
USDA Hardiness Zones	3 to 9
Native Habitat and Range	Dry or moist prairies, savannas, and open woods from Indiana and Alberta, south to Alabama and New Mexico

Purple Prairie Clover

DESCRIPTION

Purple prairie clover has 1- to 2½-inch heads of bright violet flowers in sparsely branched clusters in June and July. Each flower head consists of dozens of tiny, tightly packed flowers like those of peas. Plants form flashy, fountainlike clumps covered with blooms that are attractive to bees and butterflies. The 2- to 3-foot stems bear fine-textured leaves with three to five needle-thin lobes. The seedheads remain on the plant and are handsome in the winter landscape. Plants grow from slender, bright yellow, branching taproots that penetrate deep into the soil and enable plants to tolerate drought.

GARDEN USES

Use purple prairie clovers toward the front or middle of beds and borders in groups of three to five plants. They are well suited to prairies and meadows and make good accent plants for rock gardens. Combine them with mountain mints (*Pycnanthemum* spp.), butterfly weed (*Asclepias tuberosa*), yarrows, ornamental onions (*Allium* spp.), and smaller grasses such as little bluestem (*Schizachyrium scoparium*).

GROWING AND PROPAGATION

Plant in moist, humus-rich soil in full sun or light shade. Plants are tough and adaptable; they tolerate drought well. No serious diseases, but rabbits love to mow them down when the plants are in their prime. Clumps seldom need division. Sow seeds outdoors in autumn or indoors in June. Germination is enhanced by stratification.

OTHER RECOMMENDED *DALEA* SPECIES

Roundhead prairie clover (*Dalea multiflora*) is a more squat, bushy plant, to 2 feet tall, with dense, multiflowered clusters of white, buttonlike heads in summer. The leaves have five or more narrow leaflets. Plant in sandy or loamy, well-drained soil in full sun or light shade. Found on dry prairies, savannas, and woodland margins from Iowa and Kansas, south to Arkansas and Texas. Zones 4 to 9.

Slender white prairie clover (*Dalea candida*) is taller than purple prairie clover, growing to 4 feet, with larger, broader leaves and white flowers. The flower clusters are thicker and more elongated. Plants are often found in wetter sites and bloom a week later than purple prairie clover. Plant in sandy or loamy soil in full sun or light shade. Found on dry or moist prairies, on savannas, and in open woods from Indiana and Saskatchewan, south to Alabama and Arizona. Zones 3 to 9.

Delphinium exaltatum

Tall Larkspur

Pronunciation	dell-FIN-ee-um ex-all-TAY-tum
Family	Ranunculaceae, Buttercup Family
USDA Hardiness Zones	5 to 8
Native Habitat and Range	Rich, deciduous woods, embankments, and roadsides from Pennsylvania and Ohio, south to North Carolina and Missouri

DESCRIPTION

Tall larkspur produces erect, sparsely branching stems from 2 to 6 feet tall with deeply divided, 5-inch leaves. The ¾-inch, spurred, blue to purple flowers open for several weeks in midsummer on open-branching spikes. Each flower has five petallike sepals; the top sepal bears a long spur.

GARDEN USES

In the cultivated landscape, choose tall larkspur for informal or woodland gardens in the company of black-eyed Susans (*Rudbeckia* spp.), bergamot (*Monarda fistulosa*), Turk's-cap (*Lilium superbum*), ferns, and sedges (*Carex* spp.). In beds and borders, plant them with blazing stars (*Liatris* spp.), yarrows, garden phlox (*Phlox paniculata*), and ornamental grasses.

GROWING AND PROPAGATION

Plant in sun or light shade, in moist, alkaline to slightly acid, humus-rich soil. Top-dress clumps annually with rich compost or manure. Slugs rasp large holes in larkspur foliage. Surround your plants with a barrier strip of diatomaceous earth, or set out shallow pans of beer to trap these pests. Prevent powdery mildew, a grayish white coating on the leaves, by thinning crowded clumps to promote good air circulation. Staking tall plants helps protect the succulent stems from breaking in strong winds. Divide overgrown clumps in spring and replant in amended soil.

Tall Larkspur

Larkspurs are easily propagated by fresh seed sown immediately upon collection. Take stem cuttings in early spring from newly emerging shoots.

ANOTHER SUMMER-FLOWERING *DELPHINIUM* SPECIES

Prairie larkspur (*Delphinium virescens*) is a slender plant with 3- to 5-foot, sparsely branching stems crowned with slender spikes of 1-inch, white to pale blue flowers. The softly hairy leaves are palmate with the lobes divided into many narrow segments. The majority of the leaves are basal or on the lower half of the stems. Plant in sandy, well-drained soil in full sun or light shade. Found on dry, gravel prairies, savannas, and woodland edges from North Dakota and Illinois, south to Louisiana and Texas. Zones 3 to 9.

Dentaria laciniata

Cut-Leaved Toothwort

Pronunciation	den-TAIR-ee-uh luh-sin-ee-AH-tuh
Family	Cruciferae, Mustard Family
USDA Hardiness Zones	3 to 9
Native Habitat and Range	Moist, deciduous woods, bottomlands, streamsides, and clearings from Quebec and Minnesota, south to Florida and Oklahoma

Cut-Leaved Toothwort

DESCRIPTION

Cut-leaved toothwort produces palmately divided basal leaves with five deeply toothed leaflets. The three-stem leaves each have three deeply cut segments in a whorl below the terminal cluster of drooping, four-petaled, pale pink to white flowers. The flowers bloom in spring atop 4- to 10-inch stems; plants go dormant after flowering. Species in this genus are often included in the genus *Cardamine*.

GARDEN USES

Plant toothworts in shaded places and woodlands with other spring wildflowers such as Virginia bluebells (*Mertensia virginica*), Dutchman's breeches (*Dicentra cucullaria*), crested iris (*Iris cristata*), and spring beauties (*Claytonia* spp.). Plants form large colonies that weave through other plants. Be sure to include persistent plants such as ferns, merrybells (*Uvularia* spp.), and wild gingers (*Asarum* spp.) to fill the voids left when plants go dormant.

GROWING AND PROPAGATION

Plant in evenly moist, humus-rich soil in light to partial shade. Toothworts tolerate fairly dense shade and drier soil conditions in summer but need some direct sun in spring. Divide the plants after flowering. Sow fresh seed outdoors as soon as it ripens. Cover with ½ inch of soil.

OTHER RECOMMENDED *DENTARIA* SPECIES

Slender toothwort (*Dentaria heterophylla*) is similar to *D. diphylla* (see below), but the basal leaves are smaller and the stem leaves are narrow, lance-shaped, and sharply toothed. Plants grow 4 to 10 inches tall. Plant in rich, moist soil in light to partial shade. Found in rich, deciduous woods, wooded slopes, and floodplains from New Jersey and Indiana, south to Georgia and Mississippi. Zones 4 to 8.

Toothwort or **crinkleroot** (*Dentaria diphylla*) has broad, toothed, three-lobed leaves that form a lush but temporary groundcover. Clumps often bloom more sparsely than cut-leaved toothwort. The 6- to 12-inch bloom stalks have paired leaves below the many-flowered, pink to white inflorescences. Plant in rich, moist soil in light to partial shade. Found in low, rich, deciduous woods, bottomlands, woodland coves, and floodplains from New Brunswick and Wisconsin, south to Georgia and Alabama. Zones 4 to 8.

Dicentra cucullaria

Dutchman's Breeches

Pronunciation	dye-SEN-truh cue-cue-LAHR-ee-uh
Family	Fumariaceae, Fumitory Family
USDA Hardiness Zones	3 to 8
Native Habitat and Range	Moist, often rocky deciduous woods, bottomlands, woodland coves, and floodplains from Quebec and Minnesota, south to Georgia and Kansas

Dutchman's Breeches

DESCRIPTION

The dainty Dutchman's breeches has white flowers that resemble strings of inverted pantaloons hung out to dry in the spring breeze. The plants have 10-inch, deeply divided, ferny, blue-green leaves and 10- to 12-inch-tall bloom stalks arising from tuberous rhizomes. Colonies often form a dense carpet of foliage on the forest floor, punctuated by clusters of flowers. Plants disappear soon after flowering.

GARDEN USES

Plant Dutchman's breeches with plants that expand to fill the blank spots left when they go dormant, such as bloodroot (*Sanguinaria canadensis*), Allegheny pachysandra (*Pachysandra procumbens*), hepaticas (*Hepatica* spp.), sedges (*Carex* spp.), and ferns. Combine them under shrubs with an evergreen groundcover such as wild ginger (*Asarum* spp.). They will even grow in the shallow soil around mature tree roots because they go dormant before it becomes too dry.

GROWING AND PROPAGATION

Plant Dutchman's breeches in the shade of deciduous trees. They need spring sun and moist soil to bloom well, but shade and dry soil do them no harm once they're dormant. Sow fresh seeds outdoors; cover them with ½ inch of soil. Self-sown seedlings will appear.

ANOTHER RECOMMENDED *DICENTRA* SPECIES

Squirrel corn (*Dicentra canadensis*) is a delicate woodland plant with hyacinth-scented, white hearts crowded at the end of a weak, 8- to 10-inch stalk. They are otherwise similar in most respects to Dutchman's breeches. The ferny foliage disappears as soon as the seeds ripen in early summer. Plants reseed to form colonies that carpet the ground beneath ferns and larger wildflowers. Seedlings take several years to reach blooming size. Rodents may devour the rhizomes. Plant in moist, humus-rich soil in light to partial shade. Rich, often rocky woods, slopes, floodplains, and coves from Nova Scotia and Minnesota, south to North Carolina and Missouri. Zones 4 to 7.

Dicentra eximia

Wild Bleeding Heart, Fringed Bleeding Heart

Pronunciation	dye-SEN-truh ex-EE-me-uh
Family	Fumariaceae, Fumitory Family
USDA Hardiness Zones	3 to 9
Native Habitat and Range	Open woods, seepage slopes, and roadsides, usually in acid soils, from New Jersey and West Virginia, south to North Carolina and Tennessee, mostly in the mountains

Wild Bleeding Heart

DESCRIPTION

Wild bleeding heart is a bushy, floriferous plant with 10- to 18-inch mounds of clustered pink hearts borne throughout the spring and summer. The ferny, 12-inch, blue-green foliage sets off the flowers to good advantage. Plants grow from fibrous-rooted crowns with slow-creeping rhizomes.

GARDEN USES

Use wild bleeding heart in formal and informal plantings with garden perennials, ferns, wildflowers, and hostas. In woodlands, plant them with creeping groundcovers such as wild gingers (*Asarum* spp.), Jacob's ladder (*Polemonium reptans*), and ferns. In shade gardens, combine them with lungworts (*Pulmonaria* spp.), Siberian bugloss (*Brunnera macrophylla*), primroses, and barrenworts (*Epimedium* spp.).

GROWING AND PROPAGATION

Plant wild bleeding hearts in evenly moist, humus-rich, acidic soil in light to partial shade. In cooler climates, they can grow in full sun as long as the soil is consistently moist. Plants will bloom only in spring if they are in too much shade. The foliage remains attractive all season. Divide overgrown clumps in fall as they go dormant. Take care not to damage the brittle roots. Propagate by sowing fresh seed outdoors as soon as it is ripe. Plants self-sow.

CULTIVARS OF WILD BLEEDING HEART

'Alba' has greenish white flowers on delicate stalks.

'Boothman's Variety' has soft pink flowers and blue-gray foliage.

'Snowdrift' has pure white flowers that are smaller than the species.

Cultivars may be less vigorous than the species, depending on climate and soil.

ANOTHER *DICENTRA* SPECIES

Western bleeding heart (*Dicentra formosa*) is similar to wild bleeding heart, but its flowers are fatter and the plants spread from creeping rhizomes to form broad clumps. Many hybrids exist between the two species, and several good cultivars are available. The parentage of some cultivars is in question. 'Bountiful' has rose-red flowers and finely divided blue-green leaves. 'Luxuriant' is a vigorous, long-blooming selection with rose-red flowers and blue-gray foliage. 'Zestful' has large, deep rose-pink flowers.

Dodecatheon meadia

Shooting Star

Pronunciation	doe-deh-KATH-ee-on MEE-dee-uh
Family	Primulaceae, Primrose Family
USDA Hardiness Zones	4 to 8
Native Habitat and Range	Rich, deciduous woods, dry, rocky woods, savannas, and prairies from Maryland and Minnesota, south to Georgia and Texas

DESCRIPTION

Shooting star is an ephemeral species with 1- to 2-foot stalks of 1-inch, gracefully arching, white or pale pink flowers that resemble cyclamen with dartlike points. The thin, 1-foot, basal leaves are weakly upright to reclining, and they usually disappear after the plants bloom. 'Album' is a white selection. 'Queen Victoria' is a 10- to 12-inch plant with deep rose-purple flowers.

GARDEN USES

Interplant small groupings or drifts of shooting stars among decorative groundcovers that will hide the spaces left during dormancy. In the shade garden, choose wild gingers (*Asarum* spp.), bloodroot (*Sanguinaria canadensis*), Jacob's ladder (*Polemonium reptans*), and ferns. In prairie gardens, plant them with early wildflowers such as golden Alexanders (*Zizia* spp.), downy phlox (*Phlox pilosa*), wild geranium (*Geranium maculatum*), and violets. In a rock garden, plant them with primroses, columbines (*Aquilegia* spp.), and mounding or carpet-forming plants such as pinks (*Dianthus* spp.).

GROWING AND PROPAGATION

Plant shooting stars in moist, humus-rich soil in sun or shade. Once plants are dormant, the soil can be allowed to dry and the site can become quite shady. They prefer a neutral or slightly acidic soil. Slugs may feed on the foliage. Circle plants with a band of diatomaceous earth, or trap slugs in shallow pans of beer. Divide multicrowned clumps in summer or fall and replant the individual crowns with the fleshy, white roots spread evenly in a circle. Take root cuttings in summer or fall. Sow fresh seeds outdoors. Plants develop slowly and take several years to bloom.

Shooting Star

ANOTHER RECOMMENDED *DODECATHEON* SPECIES

Beautiful shooting star (*Dodecatheon pulchellum*) is more delicate than *D. meadia* and has 10-inch-long, oval to spatula-shaped leaves and rose-pink to deep magenta flowers. This species includes plants that were formerly classified as *D. amethystinum*. Plants vary from 3 to 20 inches tall, depending on geographic location and position. Plant in moist, humus-rich, limy soil in full sun or partial shade. Found on rock cliffs and banks and in springs and open woods from Pennsylvania and Montana, south to Arkansas and Colorado. Zones 4 to 7.

Echinacea purpurea

Purple Coneflower

Pronunciation	eck-in-AY-see-uh purr-PURR-ee-uh
Family	Asteraceae, Aster Family
USDA Hardiness Zones	3 to 8
Native Habitat and Range	Meadows, prairies, savannas, open woods, and roadsides in Ohio and Iowa, south to Georgia and Oklahoma; frequently planted along highways in other areas

Purple Coneflower

DESCRIPTION

Purple coneflower is a shrubby, branching perennial from 2 to 4 feet tall with coarse, lance-shaped leaves. Dozens of 4- to 6-inch, daisylike flowers with flat or drooping petals bloom above the foliage in summer. The petals are rose-pink to red-violet, and the orange flower centers are cone-shaped and bristly. The cones remain on the plant and are attractive in winter. Birds such as goldfinches and sparrows relish the seeds. Plants grow from fibrous taproots.

GARDEN USES

In prairie and meadow gardens, plant purple coneflower with mountain mints (*Pycnanthemum* spp.), asters, goldenrods, butterfly weed (*Asclepias tuberosa*), black-eyed Susans (*Rudbeckia* spp.), gray-headed coneflower (*Ratibida pinnata*), and grasses. Use it in beds and borders with yarrows, bee balms (*Monarda* spp.), garden phlox (*Phlox paniculata*), gayfeathers (*Liatris* spp.), and baby's-breath. Or try growing it in a deep container.

GROWING AND PROPAGATION

Plant in average to rich, loamy or sandy soil in full sun or light shade. Purple coneflower tolerates extended drought but grows best with adequate moisture. Plants increase slowly to form broad clumps. Don't divide purple coneflowers; it's seldom necessary, and the divisions tend to produce fewer flowers. To propagate, take root cuttings in fall. Sow seeds outdoors in fall or indoors in winter. Give seeds 4 to 6 weeks of cold, moist stratification to promote uniform germination. Self-sown seedlings will appear.

CULTIVARS OF PURPLE CONEFLOWER

'Bright Star' is rose-pink with mostly flat to slightly drooping flowers.

'Crimson Star' has deep crimson-purple, flat flowers on sturdy stems.

'White Lustre' has bright white flowers.

ANOTHER *ECHINACEA* SPECIES

Pale purple coneflower (*Echinacea pallida*) has stout, nearly leafless stems from 2 to 4 feet tall, topped with large heads of 4-inch, drooping, pale rose petals. The lance-shaped basal leaves are clothed in stiff hairs. Plant in average to rich soil in full sun or light shade. Found in open woods, savannas, and prairies. Zones 4 to 8.

Eryngium yuccifolium

Rattlesnake Master

Pronunciation	er-IN-gee-um yuck-ih-FOE-lee-um
Family	Apiaceae, Parsley Family
USDA Hardiness Zones	4 to 9
Native Habitat and Range	Prairies, open woods, pine and oak savannas, and roadsides in moist or dry soils from Virginia, Indiana, and Minnesota, south to Florida and Texas

DESCRIPTION

Rattlesnake master has leafy rosettes of 14-inch, lance-shaped, gray-green leaves that resemble those of yucca. Open clusters of pale green, globelike or cone-shaped flowers top thick stalks up to 3 feet tall. The flowerheads are surrounded by narrow, spiny, pale green bracts. Plants grow from a deep taproot.

GARDEN USES

In meadows and prairies, set rattlesnake master in the company of purple coneflowers (*Echinacea* spp.), goldenrods, Culver's root (*Veronicastrum virginicum*), gayfeathers (*Liatris* spp.), sneezeweeds (*Helenium* spp.), asters, pearly everlastings (*Anaphalis* spp.), and grasses. In formal gardens, contrast the bold flowers with the airy clusters of flowering spurge (*Euphorbia corollata*), baby's-breath, and sea lavender (*Limonium latifolium*).

GROWING AND PROPAGATION

Plant rattlesnake master in average to rich, moist but well-drained soil in full sun or light shade. It will thrive even in gravel and sand in full summer sun. Set out plants while they are young because they resent disturbance. If grown for too long in a container, the thick taproot will bend and the plant will not develop properly. Cut deformed taproots back to where they begin to bend. Such plants will establish more slowly but ultimately will grow better. Plants seldom need division. Sow fresh seeds outdoors in fall or indoors with 4 to 6 weeks of cold, moist stratification.

Rattlesnake Master (front)

ANOTHER *ERYNGIUM* SPECIES

Sea holly (*Eryngium aquaticum*) is similar to rattlesnake master but the leaves are broader and blunt-tipped. The flowering spheres are smaller and are surrounded by sharply toothed or spiny bracts (modified leaves). Plants grow 2 to 3 feet tall. Plant in moist to wet, sandy soil in full sun. Sea holly tolerates salt. Coastal marshes and dunes from New Jersey, south to Florida and Texas. Zones 5 to 9.

Erythronium americanum

Yellow Trout Lily

Pronunciation	air-ih-THROW-nee-um uh-mare-ih-KAH-num
Family	Liliaceae, Lily Family
USDA Hardiness Zones	3 to 9
Native Habitat and Range	Rich, deciduous woods, bottomlands, and floodplains from Nova Scotia and Minnesota, south to Florida and Alabama

Yellow Trout Lily

DESCRIPTION

Trout lilies are dainty plants with pairs of mottled leaves 4 to 8 inches long. The nodding yellow flowers have three curving petals and three petallike sepals; the sepals are tinged brown on the back. The blooms stand 6 to 12 inches high. The plants spread by slender runners that form bulbs at their ends, creating large colonies. The plants go dormant immediately after flowering. This species seems to have two distinct forms: one is large, with flowers bearing yellow stamens; the other is small, with flowers bearing brown stamens.

GARDEN USES

Trout lilies are perfect companions for spring bulbs and early perennials. Combine them with wildflowers such as Dutchman's breeches (*Dicentra cucullaria*), spring beauties (*Claytonia* spp.), creeping Jacob's ladder (*Polemonium reptans*), and bloodroot (*Sanguinaria canadensis*). Or, plant them among ferns and foliage plants such as wild ginger (*Asarum* spp.), lungworts (*Pulmonaria* spp.), wild bleeding heart (*Dicentra eximia*), and hostas, which will fill the gaps left when trout lilies go dormant in summer.

GROWING AND PROPAGATION

Plant in moist, humus-rich soil. Spring sun is essential, but once the plants are dormant, the site can become quite shady. There are often many more leaves than flowers. Take care not to dig into the clumps during the dormant season. Sow fresh seed outdoors as soon as it is ripe. Seedlings develop slowly and may take 3 to 5 years to bloom.

ANOTHER *ERYTHRONIUM* SPECIES

White trout lily (*Erythronium albidum*) is similar to yellow trout lily but has white flowers with a purple or blue blush on the outside of the sepals. Plant in moist, rich soil in sun or shade. Found in low, deciduous woods, rocky slopes, bottomlands, and floodplains in limy soils from Ontario and Minnesota, south to Virginia, Kentucky, and Texas. Zones 4 to 8.

Eupatorium coelestinum

Hardy Ageratum

Pronunciation	you-puh-TOR-ee-um see-less-TIE-num
Family	Asteraceae, Aster Family
USDA Hardiness Zones	4 to 10
Native Habitat and Range	Open woods, low meadows, streamsides, and ditches from New Jersey and Illinois, south to Florida and Texas

Hardy Ageratum

DESCRIPTION

Hardy ageratum is an open, bushy plant that looks like a tall version of annual ageratum and spreads rapidly from creeping rhizomes. The sprawling stems grow to 2 to 3 feet, topped with small clusters of powder blue flowers in late summer and fall.

GARDEN USES

Hardy ageratum is usually too aggressive for borders, but when planted within a barrier, it adds a lovely blue accent to the late-season garden. Plant a clump in a large pot with the bottom cut out. Sink the pot in the garden up to within 1 inch of the rim. Combine it with 'Moonbeam' coreopsis, smaller goldenrods, and the foliage of ligularias and hostas. Hardy ageratum is best suited to informal settings where its lightning-fast spread will not compromise an intricate design. In meadows and along roadside ditches and driveways, combine it with other late-season bloomers such as fall asters, goldenrods, marsh mallows (*Hibiscus* spp.), sneezeweed (*Helenium autumnale*), and grasses.

GROWING AND PROPAGATION

Plant in rich, moist soil in full sun or light shade. Dig out sections around the edges of clumps freqently to control their spread. Propagate by division in spring or tip cuttings taken in early summer.

OTHER FAST-SPREADING *EUPATORIUM* SPECIES

Boneset (*Eupatorium perfoliatum*) is a delicate, white-flowered species with 6-inch, opposite, lance-shaped foliage joined in the middle and pierced by the sturdy 3- to 5-foot stem. Plants bloom in mid- to late summer. Plant in rich, consistently moist soil. Found in wet areas from Nova Scotia and Quebec, south to Florida and Oklahoma. Zones 3 to 8.

Joe-Pye weed (*Eupatorium fistulosum*) is a giant among Joe-Pyes, with straight stems that may reach 14 feet tall, although 6-foot stems are more typical. This stately plant has four or five 10- to 14-inch leaves per whorl and elongated domes of dusty rose flowers in summer. Plant in average to rich, moist soil in full sun or light shade. Found in meadows and ditches from Maine and Iowa, south to Florida and Texas. 'Bartered Bride' is a good white selection. Zones 4 to 9.

White snakeroot (*Eupatorium rugosum*) has opposite, 4-inch, triangular leaves and white flowers in small clusters on 3- to 4-foot stalks. The silver seedheads are very showy. The plant is poisonous. White snakeroot thrives in the dry shade of woodlands. Its late-season white flowers brighten dark areas. Found in open woods and on roadsides from Nova Scotia and Saskatchewan, south to Georgia and Texas. Zones 3 to 8.

Eupatorium purpureum

Joe-Pye Weed

Pronunciation	you-puh-TOR-ee-um purr-PURR-ee-um
Family	Asteraceae, Aster Family
USDA Hardiness Zones	3 to 8
Native Habitat and Range	Wet slopes, low meadows, woodland edges, and roadsides from New Hampshire and Iowa, south in the mountains to Georgia and west to Oklahoma

Joe-Pye Weed

DESCRIPTION

Joe-Pye weed forms 3- to 6-foot clumps crowned by mounded or domed clusters of pale rose or light purple, sweet-scented flowers. Individual flowers are small and fuzzy. The leaves circle the stems in whorls of four; leaves may grow to 12 inches long. The seedheads are attractive in winter. Plants arise from thick crowns.

GARDEN USES

In meadows and other informal sites, plant Joe-Pye weed with asters, bee balms (*Monarda* spp.), lilies, tall larkspur (*Delphinium exaltatum*), goldenrods, and grasses. Use the large clumps and soft, cottony flowers to screen views and provide a background to beds and borders. Combine it with sneezeweed (*Helenium autumnale*), rose mallow (*Hibiscus moscheutos*), purple coneflowers (*Echinacea* spp.), garden phlox (*Phlox paniculata*), daisies, and hardy mums.

GROWING AND PROPAGATION

Plant in moist, average to rich soil in full sun or light shade. Joe-Pye weed is an easy-care plant that needs little attention once established. It takes at least two seasons for new plants to reach full size. Divide oversize clumps in spring or fall. Separate the tough crown into sections using a sharp knife or shears. Replant in well-prepared soil. Take stem cuttings in early summer. Sow seeds outdoors or stratify for 4 weeks before sowing indoors.

To lessen occasional powdery mildew problems, thin clumps to improve air circulation. Leafminers form tunnels in the leaves. These generally don't harm the plant; handpick damaged leaves.

RECOMMENDED CULTIVARS OF JOE-PYE WEED

'Atropurpureum' is a compact plant that grows up to 5 or 6 feet tall with deep purple stems, dark leaves, and purple, sweet-scented flowers. This selection may be a hybrid with *E. maculatum*.

'Big Umbrella' has dark stems and huge heads of dark raspberry flowers.

'Gateway' grows at least 6 feet tall and has pale mauve flowers in basketball-size clusters.

Euphorbia corollata

Flowering Spurge

Pronunciation	you-FOR-bee-uh core-oh-LAY-tuh
Family	Euphorbiaceae, Spurge Family
USDA Hardiness Zones	3 to 9
Native Habitat and Range	Open woods, savannas, meadows, prairies, and roadsides from Massachusetts and Minnesota, south to Florida and Texas

Flowering Spurge

DESCRIPTION

Flowering spurge is a creeping plant with slender, 1- to 3-foot stems that bleed milky sap when picked or damaged. Pale green, 3-inch leaves sparsely cover the stems. Broad clusters of flowers with pure white bracts (modified leaves) top the plant, making it look like a tall, sturdy baby's-breath.

GARDEN USES

Choose flowering spurge for informal gardens as well as meadow and prairie plantings. Combine the airy heads with purple coneflowers (*Echinacea* spp.), bee balms (*Monarda* spp.), blazing stars (*Liatris* spp.), sunflower heliopsis (*Heliopsis helianthoides*), Joe-Pye weeds (*Eupatorium* spp.), ironweeds (*Vernonia* spp.), goldenrods, and asters. In formal beds and borders, plant flowering spurge in the gaps between other perennials to tie the planting together. Plant them with cannas, sea hollies (*Eryngium* spp.), sages (*Salvia* spp.), phlox, and ornamental grasses. Plants grow well on dry clay banks and roadsides where few other plants thrive. They also are well suited to container growing.

GROWING AND PROPAGATION

Plant in average to rich, well-drained soil in full sun or partial shade. Flowering spurge is a long-lived perennial that tolerates drought and needs little care. Divide clumps as needed to control their spread or for propagation. Propagate by taking tip cuttings in summer. Be sure to stick cuttings in well-drained soil before the stem end dries out.

OTHER *EUPHORBIA* SPECIES

Leafy spurge (*Euphorbia esculenta*) is a fast-growing, invasive species that is considered a noxious weed in many states. Watch for its attractive lacy foliage and yellow flowers in your garden or meadow. Do not cultivate this plant; dig it out wherever it occurs.

Snow-on-the-mountain (*Euphorbia marginata*) is a 1- to 3-foot annual with oval leaves topped by broad, dense clusters of showy oval bracts (modified leaves) outlined in white. Plant in poor to average well-drained soil in full sun. Plants will self-sow. Found on prairies, high plains, waste places, and roadsides from Minnesota and Montana, south to Missouri and New Mexico. Zones 3 to 9.

Filipendula rubra

Queen-of-the-Prairie

Pronunciation	fill-uh-PEN-djew-luh ROO-bruh
Family	Rosaceae, Rose Family
USDA Hardiness Zones	3 to 9
Native Habitat and Range	Low meadows and prairies, wetlands, and ditches from New York and Wisconsin, south to North Carolina and Kentucky (has naturalized outside its range)

Queen-of-the-Prairie

DESCRIPTION

Queen-of-the-prairie is a stately plant that grows to 6 feet tall and has bold, deep green, divided leaves up to 1 foot long. The flowers resemble hot-pink or rose cotton candy and may reach 9 inches across. The seedheads are attractive in fall and winter. 'Venusta' and 'Venusta Magnifica' have deep pink flowers on 4- to 5-foot plants.

GARDEN USES

Use queen-of-the-prairie in formal borders or in meadow and pondside plantings. Pair them with shrubs or combine them with ferns, irises, bee balms (*Monarda* spp.), Joe-Pye weeds (*Eupatorium* spp.), and grasses. In formal beds, plant them with bellflowers (*Campanula* spp.), purple coneflower (*Echinacea purpurea*), lupines (*Lupinus* spp.), phlox, and delphiniums.

GROWING AND PROPAGATION

Plant queen-of-the-prairie in evenly moist, humus-rich soil in full sun or light shade. They will thrive in highly moist soils and beside creeks or ponds. If leaves become tattered or crispy, cut them to the ground; fresh foliage will emerge. Clumps spread rapidly by creeping stems and need frequent division to keep them from overrunning other plants. Use a knife or shears to cut through tough clumps.

Powdery mildew and spider mites are occasional problems on plants grown in hot, dry places. Keep plants well watered to prevent both these problems. For powdery mildew, thin clumps to promote better air circulation.

A PERFECT GARDEN COMPANION

Wild quinine (*Parthenium integrifolium*) is an attractive plant with deep green, toothed, broad, lance-shaped leaves on stout stems. Plants produce flat clusters of knobby white flowers in mid- to late summer. Plants thrive in moist to wet prairies, often alongside queen-of-the-prairie. They make good garden companions because the wild quinine flowers open as queen-of-the-prairie's flowers fade, thereby extending the season. The dark foliage is a nice addition to the late-season garden with asters and goldenrods. Zones 4 to 8.

Gaillardia pulchella

Blanket Flower

Pronunciation	gah-LARD-ee-uh pull-CHELL-luh
Family	Asteraceae, Aster Family
USDA Hardiness Zones	4 to 9
Native Habitat and Range	Open woods, meadows, plains, and dunes from Missouri and Colorado, south to Mexico, and along the coast from Virginia to Texas (has naturalized outside its range)

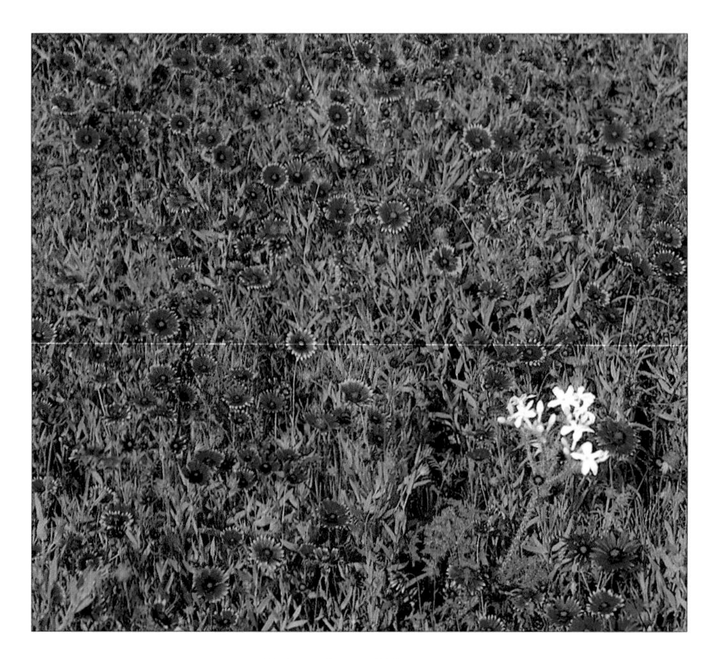

Blanket Flower

DESCRIPTION

This showy annual or short-lived perennial flowers throughout summer with ragged yellow and orange, daisy-like flowers on loose, 2- to 3-foot stems. The hairy, lobed basal leaves are 8 to 10 inches long; stem leaves are smaller. The spiny, globe-shaped seedheads remain attractive after the flowers fade.

GARDEN USES

Blanket flowers produce mounds of brilliant flowers from early summer through fall. They are perfect for borders, rock gardens, seaside gardens, and containers. Plant them with other warm-colored perennials such as coreopsis, butterfly weed (*Asclepias tuberosa*), and goldenrods. In formal gardens, plant them with yarrows, yuccas (*Yucca* spp.), and salvias.

GROWING AND PROPAGATION

Plant blanket flower in average, well-drained, sandy or loamy soil in full sun. They thrive on heat and benign neglect, and they are perfect for seaside gardens because they withstand salt and drought. Plants tend to die out in 1 to 2 years but they reseed freely. For longer-lived plants, choose one of the species listed at right. Propagate by stem cuttings in early summer. Sow seeds outdoors in fall or indoors after 4 weeks of cold, moist stratification. Seedlings develop quickly and bloom the first year.

Blanket flower leaves are prone to powdery mildew. To prevent problems, keep plants well watered and thin stems to promote air circulation.

OTHER *GAILLARDIA* SPECIES

Blanket flower (*Gaillardia aristata*) is a perennial with yellow flowers that are often blazed with brown. Flowers have purple-brown centers. They grow 1 to 2½ feet tall. Plant in average, well-drained soil in full sun. Dry prairies, high plains, and roadsides from Saskatchewan and British Columbia, south to New Mexico and Oregon. Zones 2 to 10.

Hybrid blanket flower (*Gaillardia* × *grandiflora*) is a showy cross between *G. aristata* and *G. pulchella*. This popular plant is floriferous and hardy but may be short-lived because of its annual parentage. The flowers are orange and yellow, often with brick red bands or eyes. Plant in average, well-drained soil in full sun or light shade. 'Baby Cole' is a floriferous, long-lived dwarf selection only 8 inches tall. 'Bremen' has copper-red flowers tipped in yellow on 2- to 3-foot stems. 'Dazzler' has yellow-centered flowers with red tips. 'Goblin' is a 12-inch dwarf with red and yellow flowers. Zones 4 to 9.

Gaultheria procumbens

Wintergreen, Teaberry

Pronunciation	gall-THEER-ee-uh proh-KUM-behnz
Family	Ericaceae, Heath Family
USDA Hardiness Zones	3 to 8
Native Habitat and Range	Dry or moist woods, outcroppings, and forested bogs from Newfoundland and Manitoba, south to Virginia and Minnesota, and in the mountains of North Carolina

DESCRIPTION

Wintergreens are woody groundcovers with trailing stems and creeping rhizomes. They form broad open colonies as they mature. The oval, 1- to 2-inch shining evergreen leaves are crowded at the top of a 2- to 6-inch stalk. The sweet-scented leaves can be dried and used to make a pungent tea with the familiar flavor of wintergreen. The ¼-inch, bright red berries are similarly fragrant and edible. Birds adore the berries, but if they're overlooked in the autumn, the fruits will persist through the winter and into the next season.

GARDEN USES

Wintergreen grows well under shrubs and flowering trees in acid soil. Combine it with galax (*Galax rotundifolia*), round-lobed hepatica (*Hepatica americana*), vernal iris (*Iris verna*), Canada mayflower (*Maianthemum canadense*), and wild ginger (*Asarum* spp.). Or, try growing it in a large container with a dwarf shrub.

GROWING AND PROPAGATION

Plant in moist or dry, humus-rich, acid soil in sun or shade. Plants are slow to establish, so be sure to water well the first year. Divide the rhizomes in the spring and treat them like cuttings until well rooted.

Wintergreen

ANOTHER *GAULTHERIA* SPECIES FOR SHADE

Creeping snowberry (*Gaultheria hispidula*) is a dainty trailing plant with tiny, oval, evergreen leaves and snow white berries in autumn and winter. Plants form a tight, fine-textured groundcover on mossy logs and in rich, humusy soil. They creep close to the ground among ferns, starflower (*Trientalis borealis*), and acid-loving shrubs. Plant in cool, moist, humus-rich, acid soil in partial to full shade. Forested bogs and damp, coniferous woods and on mossy logs from Labrador and British Columbia, south to the mountains of North Carolina and to Minnesota and Oregon. Zones 3 to 7.

A PERFECT GARDEN COMPANION

Starflower (*Trientalis borealis*) bears ½-inch, snow white flowers with seven pointed petals in late spring and summer. Five to ten bright green, lance-shaped leaves form a whorl on a thin stalk to 9 inches tall. Plants spread by thin rhizomes with fibrous roots. Use them with wintergreens near woodland garden paths where you can enjoy the delicate flowers. Plant in humus-rich, moist or dry, acidic soil in light to deep shade. Divide plants after flowering or in early fall. Deciduous or coniferous woods and bogs. Zones 2 to 8.

Gaura lindheimeri

White Gaura

Pronunciation	GAW-ruh lind-HIGH-mer-eye
Family	Onagraceae, Evening Primrose Family
USDA Hardiness Zones	5 to 9
Native Habitat and Range	Open low woods, prairies, pinelands, and roadsides in Louisiana and Texas

White Gaura

DESCRIPTION

White gaura is an airy, shrubby perennial with erect, 3- to 4-foot stems and deep green leaves. The stems bear spikes of 1-inch, white, four-petaled flowers above the foliage. The flowers turn pale rose as they age. In warm regions, the plant blooms from June through at least September. In northern areas, flowers appear mainly in late summer and fall. Clumps grow from a thick, deep taproot.

GARDEN USES

Choose white gaura for borders or meadow gardens or as an accent plant. Combine it with ornamental grasses and small-flowered plants such as coreopsis, verbenas (*Verbena* spp.), sedums, sea lavender (*Limonium latifolium*), and hardy cranesbills (*Geranium* spp). Use it to tie together bold meadow plants such as gayfeathers (*Liatris* spp.), Joe-Pye weeds (*Eupatorium* spp.), purple coneflowers (*Echinacea* spp.), asters, phlox, and sunflowers.

GROWING AND PROPAGATION

Plant in average to rich, moist, well-drained soil in full sun. Established plants will tolerate drought but perform best with even moisture. Gardeners in warm zones prize gauras because they bloom well even in hot, humid conditions. Plants seldom need division. Sow seeds outside in fall. Self-sown seedlings will likely appear.

ANOTHER *GAURA* SPECIES FOR A SUNNY SITE

Scarlet gaura (*Gaura coccinea*) has 2-foot stems densely covered by 1-inch, narrow leaves. The ¼-inch rose-pink to red flowers crowd at the tips of the stems. Plant in average to rich, well-drained soil in full sun. Found on dry prairie ridges, savannas, plains, and roadsides from Indiana and Alberta, south to Missouri, Texas, and California. Zones 3 to 9.

A PERFECT GARDEN COMPANION

Blackfoot daisy (*Melampodium leucanthum*) is covered with 1-inch, yellow-centered white daisies for at least four months in spring and summer. Mature plants form dense, attractive mounds from tough, branched taproots. When out of bloom, the soft, gray-green foliage is an attractive foil for other flowering plants. Use them in borders, informal gardens, or containers as a mounding groundcover under clumps of white gaura. Both plants will bloom the whole season. Plant in average to lean, well-drained, neutral to slightly acid soils in full sun. Plants are tough and adaptable. Found on dry, gravelly, limestone prairies and slopes. Zones 7 to 9.

Gentiana andrewsii

Bottle Gentian

Pronunciation	jen-SHE-ah-nuh an-DROOZ-ee-eye
Family	Gentianaceae, Gentian Family
USDA Hardiness Zones	3 to 8
Native Habitat and Range	Moist, open woods, wet meadows, marshes, and ditches from Quebec and Manitoba, south to New Jersey, in the mountains to North Carolina, and west to Nebraska

Bottle Gentian

DESCRIPTION

Bottle gentian has erect to sprawling 1½- to 2-foot stems with glossy, oval to broadly lancelike, 4-inch leaves. The 1-inch, bottlelike, closed flowers are an intense indigo blue called gentian blue. Inside the closed tip are five fringed lobes. Plants bloom from late summer through fall.

GARDEN USES

Combine bottle gentian with ferns, grasses, turtleheads (*Chelone* spp.), goldenrods, and asters in the late-season garden. Their rich blue color is lovely with small-leaved hostas, toad lilies (*Tricyrtis* spp.), variegated sedges (*Carex* spp.), and other lush foliage plants.

GROWING AND PROPAGATION

Plant bottle gentian in evenly moist, humus-rich soil in full sun or partial shade. Provide shade from hot afternoon sun to avoid leaf scald, especially in warmer zones. Bottle gentian is long-lived and thrives with little care. Plants seldom need division and dislike root disturbance. Propagate by lifting clumps in spring and carefully splitting the crowns. Replant the divisions immediately. Sow fresh seeds in late winter with 4 to 6 weeks of cold, moist stratification. Plants take up to 3 years to bloom from seed.

OTHER *GENTIANA* SPECIES

Closed gentian (*Gentiana clausa*) is nearly identical to bottle gentian, but the lobes of the petals are less fringed. Plant in moist, rich soil in sun or partial shade. Plants grow naturally in moist open woods, low meadows, and ditches from Maine and New York, south to New Jersey, and in the mountains to North Carolina. Zones 4 to 8.

Fringed gentian (*Gentiana crinita,* also listed as *Gentianopsis crinita*) is a biennial with stunning, funnel-shaped flowers with five rounded, fringed petals of steely purple. This is one of the few eastern gentians with flowers that are actually open. The flowers close up in the evening. Plants have pairs of narrow, shiny green leaves on stiff, 3-foot stems. Plants must be grown from seed each year unless they self-sow, which is rare. Plant in moist, rich neutral soil in full sun or light shade. Found in low meadows and prairies and along roadsides in soils that are high in manganese from Maine to Manitoba, south to Georgia and Iowa. Zones 4 to 8.

Sampson's snakeroot (*Gentiana villosa*) grows to 2½ feet and has deep green leaves and green flowers on stems that spread to form an open clump. Plant in moist, humus-rich soil in light to full shade. Found in open, deciduous woods, and bottomlands and along shaded roadsides from New Jersey and Indiana, south to Florida and Louisiana. Zones 5 to 9.

Geranium maculatum

Wild Geranium, Wild Cranesbill, Spotted Cranesbill

Pronunciation	jer-ANE-ee-um mack-you-LAY-tum
Family	Geraniaceae, Geranium Family
USDA Hardiness Zones	3 to 8
Native Habitat and Range	Open deciduous woods, bottomlands, prairies, savannas, and roadsides from Maine and Manitoba, south to Georgia and Arkansas

Wild Geranium

DESCRIPTION

Geranium species are called cranesbills because the seedpods have a rigid, tapered "beak" that serves as a launching pad to project the seeds outward from the plants. Wild geranium is a 1- to 2-foot plant that forms open clumps of rounded, deeply lobed leaves. The five-petaled, rose-pink to white flowers bloom in loose clusters in spring. The foliage turns a lovely burgundy-red, scarlet, or orange in the fall. 'Alba' is the name given to white and pale pink selections grown from seed. 'Hazel Gallagher' is a dependable cultivar with large, pure white flowers.

GARDEN USES

Plant wild geranium in informal woodland gardens, prairie gardens, and formal perennial beds. In woodland and prairie gardens, combine them with lupines (*Lupinus* spp.), golden Alexanders (*Zizia* spp.), wild blue phlox (*Phlox divaricata*), gentians (*Gentiana* spp.), ferns, and grasses.

GROWING AND PROPAGATION

Plant in moist, well-drained, humus-rich soil in full sun or shade. Plants bloom more freely with some direct sun. Divide the slow-creeping rhizomes in early spring, after flowering, or in the fall. Sow seeds outdoors or inside on a warm seedbed. Seedlings develop in 3 to 5 weeks.

OTHER *GERANIUM* SPECIES

Richardson's geranium (*Geranium richardsonii*) has deeply cut, three- to five-lobed leaves and open clusters of 1-inch, white to pink flowers with rose-pink veins. Plants bloom in early to midsummer. The foliage turns burgundy and purple in the fall. Plant in moist, well-drained, humus-rich soil in full sun or partial shade. Open woods, meadows, and prairies in moist soil from Saskatchewan and British Columbia, south to New Mexico and California. Zones 4 to 8.

Sticky cranesbill (*Geranium viscosissimum*) is similar to *G. richardsonii* but has glandular hairs on the bloom stalk and flowers. Plants grow to 2 feet tall. Plant in moist or dry, rich soil in full sun to partial shade. Open, mixed woods, meadows, and streamsides from Saskatchewan and British Columbia, south to Nevada and California. Zones 4 to 8.

Geum triflorum

Prairie Smoke

Pronunciation	JEE-um try-FLOOR-um
Family	Rosaceae, Rose Family
USDA Hardiness Zones	1 to 8
Native Habitat and Range	Dry to moist gravel prairies or black soil prairies, savannas, and high mountain meadows from the Canadian tundra, south to New York, through the Great Lakes states, and across the Plains to the southern Rockies and the Sierras

Prairie Smoke

DESCRIPTION

Prairie smoke has a basal rosette of divided leaves crowned by furry 6- to 12-inch bloom stalks that bear flowers in groups of three in spring and early summer. The nodding, rose-pink flowers have long purple bracts (modified leaves). The flowers fade to reveal dense heads of feathery, pale rose-pink plumes that look like smoke.

GARDEN USES

Prairie smoke quickly forms a dense groundcover from woody rhizomes. It is a perfect choices for prairie and meadow plantings, rock gardens, or the front of beds and borders. Plant it alone or in combination with grasses, spring bulbs, and low perennials such as creeping phlox (*Phlox stolonifera*), hardy cranesbills (*Geranium* spp.), green and gold (*Chrysogonum virginianum*), and sedums.

GROWING AND PROPAGATION

Plant in average, well-drained soil in full sun or partial shade. Prairie smoke survives bitter cold, high heat, poor soil, and prolonged drought. It performs best in northern areas with cool summers. If you plan to plant prairie smoke near the extremes of its range, buy local stock to ensure good performance. Lift and divide clumps in early spring or fall. Store dry seed until spring, or sow immediately in an outdoor seedbed. Indoor sowing requires cold, moist stratification. First-year seedlings develop slowly. Transplant when they are 2 to 3 inches high.

A *GEUM* SPECIES FOR WET GARDENS

Purple avens (*Geum rivale*) has tall stalks bearing nodding, purple, bell-like flowers in groups of three. After flowers fade, it has tufted plumes of seeds. Plant in moist to wet, humus-rich soil in full sun or partial shade. Bogs, low woods, and wet meadows from Newfoundland and Alberta, south to New Jersey, Michigan, and California. 'Album' has white flowers. 'Leonard's Variety' is a hybrid with wine-red flowers. Zones 2 to 7.

Gillenia trifoliata (also listed as *Porteranthus trifoliatus*)

Bowman's-Root

Pronunciation	gih-LEN-ee-uh tri-foe-lee-AH-tuh
Family	Rosaceae, Rose Family
USDA Hardiness Zones	4 to 8
Native Habitat and Range	Open woods, clearings, rocky slopes, and roadsides from southern Ontario and New England, south to Georgia and Alabama

Bowman's-Root

DESCRIPTION

Bowman's-root is an erect, shrublike perennial with wiry stems sparsely covered by toothed leaves with three leaflets. Plants reach from 2 to 4 feet tall and wide and have thick, deep roots. Broad terminal clusters of four-petaled, starry white or pinkish flowers smother the plant in late spring or early summer. The leaves turn rich yellow in autumn. The plant is poisonous.

GARDEN USES

Mimic bowman's-root's natural habit by planting it in large patches along with coreopsis, fire pink (*Silene virginica*), thick leaf phlox (*Phlox carolina*), and grasses. Or plant it at the edge of a woodland with ferns, irises, bee balms (*Monarda* spp.), baptisias (*Baptisia* spp.), and Culver's root (*Veronicastrum virginicum*). In beds and borders, combine it with garden phlox (*Phlox paniculata*), lupines (*Lupinus* spp.), hardy cranesbills (*Geranium* spp.), and artemisias (*Artemisia* spp.).

GROWING AND PROPAGATION

Plant in rich, moist soil in sun or partial shade. Established plants are drought-tolerant. Bowman's-root will tolerate full sun in all but the hottest southern regions. Plants spread slowly to form tight clumps that seldom need dividing. Take stem cuttings in spring, or sow seed outdoors when ripe in late summer. Plants take 3 to 4 years to attain good form.

ANOTHER *GILLENIA* SPECIES

American ipecac (*Gillenia stipulata*) is similar to bowman's-root, but the plants are more open and the stems bear persistent, leaflike structures called stipules where the leaves join the stem, giving the plant a winged appearance. Plant in average to rich, moist soil in full sun or partial shade. Found in open woods and clearings and along roadsides from New York and Illinois, south to Georgia and Texas. Zones 4 to 9.

A PERFECT GARDEN COMPANION

Wood mint (*Blephilia ciliata*) forms erect, 1- to 1½-foot clumps crowned by whorled tiers of tightly packed, pink to violet-blue flowerheads. The 3-inch, oval leaves are softly hairy and turn a decorative shade of rose-pink in late summer. Plant them in groups at the edge of a woodland or along a path with bowman's-root and other open, airy plants. Plants have an upright character and in formal garden situations, they form large showy clumps. Plant in average to rich, moist but well-drained soil in light to partial shade. Plants spread slowly and division is rarely necessary. Found in woods and clearings and along roadsides. Zones 4 to 8.

Helenium autumnale

Common Sneezeweed

Pronunciation	hel-EE-nee-um awe-tum-NAH-lee
Family	Asteraceae, Aster Family
USDA Hardiness Zones	3 to 8
Native Habitat and Range	Low woods, wet meadows and prairies, marshes, and ditches from Quebec and British Columbia, south to Florida and Arizona

Common Sneezeweed

DESCRIPTION

Common sneezeweed has bright green, lancelike leaves with toothed edges that arise from a thick crown. It grows 3 to 5 feet tall, and the stems are topped by broad clusters of yellow or orange, daisylike, 2-inch flowers with yellow centers. Plants in moist, rich sites tend to be quite robust while those in dry soil are short and delicate.

GARDEN USES

Sneezeweed adds a bright splash of color to late-season borders and informal meadow plantings. In meadows or beside ponds, combine it with irises, ferns, ornamental grasses, hibiscus, New England aster (*Aster novae-angliae*), Joe-Pye weeds (*Eupatorium* spp.), and ironweeds (*Vernonia* spp.). In beds and borders, plant it with butterfly bushes (*Buddleia* spp.), garden phlox (*Phlox paniculata*), asters, coreopsis, and goldenrods.

GROWING AND PROPAGATION

Plant in evenly moist, humus-rich soil in full sun or light shade. Plants will also grow in wet soil. In warmer climates, plants tend to stretch; pinch them in the spring to promote compact growth. Some cultivars are naturally compact and self-supporting. Divide clumps every 3 to 4 years to keep them vigorous. Propagate by stem cuttings in early summer. Sow seeds outside in fall; cultivars will not come true from seed.

RECOMMENDED CULTIVARS OF COMMON SNEEZEWEED

'Brilliant' is a 3-foot plant with orange-bronze, dark-centered flowers.

'Butterpat' is a 3- to 4-foot plant with bright yellow flowers.

'Crimson Beauty' has mahogany flowers.

'Riverton Beauty' has golden yellow flowers with bronze-red centers.

ANOTHER *HELENIUM* SPECIES

Purple-headed sneezeweed (*Helenium flexuosum*) is a compact grower to 3 feet with leafy stems and broad, domed clusters of flowers with drooping yellow petals and brownish purple spherical centers. Plants bloom from mid- to late summer into early autumn. Plant in moist or dry, rich soil in full sun or light shade. Plants are adaptable and will self-sow in the garden. Found in low open woods, pine barrens, and waste places and along roadsides from Massachusetts and Illinois, south to Florida and Texas. Zones 5 to 9.

Helianthus angustifolius

Swamp Sunflower

Pronunciation	hee-lee-AN-thus an-gus-tih-FOE-lee-us
Family	Asteraceae, Aster Family
USDA Hardiness Zones	6 to 9; hardy farther north but does not bloom before frost
Native Habitat and Range	Low woods, streamsides, and ditches from New York and Indiana, south to Florida and Texas, mostly along the coastal plain and the banks of the lower Mississippi River

Swamp Sunflower

DESCRIPTION

Swamp sunflower has thick, 4- to 10-foot stems covered with deep green, lancelike leaves that can grow to 1 foot long. Branching clusters of 3-inch, bright yellow, daisylike flowers top the stems in September and October. The flowers have purple centers.

GARDEN USES

Plant swamp sunflower at the back of a large perennial garden. It offers a great end-of-season show when combined with fall-blooming asters, sneezeweeds (*Helenium* spp.), goldenrods, and ornamental grasses. Use swamp sunflower and other *Helianthus* species in borders, informal meadows, and prairie plantings with ironweeds (*Vernonia* spp.), yarrows, gayfeathers (*Liatris* spp.), purple coneflowers (*Echinacea* spp.), wild bergamot (*Monarda fistulosa*), milkweeds (*Asclepias* spp.), and ornamental grasses.

GROWING AND PROPAGATION

Plant in moist, average to rich soil in full sun. All *Helianthus* species are drought-tolerant once established but also tolerate wet soil. Swamp sunflower is one of the largest wildflowers, surpassed in size only by Joe-Pye weed and a few others. Give it ample room to grow to mature height and spread. The erect, stiff stems seldom need staking except in windy areas. Divide overgrown clumps every 3 to 4 years in the spring. Propagate by stem cuttings in early summer or by seed sown outdoors when ripe.

MORE *HELIANTHUS* SPECIES FOR FULL SUN

Maximilian sunflower (*Helianthus maximiliani*) has gray-green, drooping, lancelike leaves that grow up to 1 foot long. The long flower clusters are similar to those of swamp sunflower. Plants reach 4 to 8 feet tall, occasionally taller. Plant in moist, rich soil in full sun. Found in low meadows, low prairies, and ditches from Minnesota and British Columbia, south to Missouri, Texas, and Wyoming; naturalized in the east. Zones 3 to 8.

Willow leaf sunflower (*Helianthus salicifolius*) is grown as much for its narrow, 8-inch, drooping gray-green leaves as for its flowers. The slender upright clusters of 2-inch flowers are showy throughout autumn. Plant in average, sandy or loamy, well-drained soil in full sun. Found on dry prairies, savannas, and roadsides in limy soil from Illinois and Missouri, west to Kansas and Oklahoma. Zones 4 to 8.

Heliopsis helianthoides

Sunflower Heliopsis, Oxeye

Pronunciation	hee-lee-OP-sis hee-lee-an-THOY-deez
Family	Asteraceae, Aster Family
USDA Hardiness Zones	3 to 9
Native Habitat and Range	Open woods, prairies, meadows, and plains from Quebec and British Columbia, south to Georgia and New Mexico

Sunflower Heliopsis

DESCRIPTION

Sunflower heliopsis looks like a well-branched sunflower. The bright yellow, 2- to 3-inch flowers bloom throughout summer. The plants are bushy, reach 3 to 6 feet, and have triangular leaves borne in pairs on the slender stems.

GARDEN USES

Sunflower heliopsis is an enchanting plant for meadow and prairie gardens. Its early and persistent blooms complement downy phlox (*Phlox pilosa*), gayfeathers (*Liatris* spp.), bee balms (*Monarda* spp.), baptisias (*Baptisia* spp.), and milkweeds (*Asclepias* spp.). In beds and borders combine it with garden phlox (*Phlox paniculata*), hibiscus, hardy mums (*Chrysanthemum* spp.), speedwells (*Veronica* spp.), asters, and ornamental grasses.

GROWING AND PROPAGATION

Plant in moist or dry, average to rich soil in full sun or partial shade. Sunflower heliopsis grows best in borders in full sun. Most named cultivars have self-supporting stems. The wild forms are less rigid in habit. Clumps will bloom for 4 to 6 weeks in summer. Divide the crowns every 2 to 3 years to keep the plants vigorous. Cultivars are more vigorous; divide them only when necessary to control their spread or to rejuvenate the clumps. Propagate by stem cuttings taken in late spring. Self-sown seedlings may be plentiful.

The leaves may show the white, powdery covering characteristic of powdery mildew, but this should not harm the plants overall.

RECOMMENDED SELECTIONS OF SUNFLOWER HELIOPSIS

'Golden Plume' is a double-flowered, compact selection growing 3 to 3½ feet tall.

'Hohlspiegel' has semidouble flowers with 4-foot stems.

'Incomparabilis' has yellow-orange, 3-inch, semidouble flowers.

'Karat' has 3-inch, bright yellow, single flowers on 4- to 5-foot plants.

'Patula' has glossy, semidouble flowers on 3½-foot stems and an extended bloom season.

'Summer Sun' is a heat-tolerant, compact grower (to 3 feet) with 4-inch single flowers.

H. helianthoides* var. *scabra is a thick plant with large, rough foliage and plentiful flowers; 'Gold Greenheart' has double yellow-green flowers with green centers.

Hepatica americana

Round-Lobed Hepatica

Pronunciation	heh-PAT-ih-cuh uh-mare-ih-KAH-nuh
Family	Ranunculaceae, Buttercup Family
USDA Hardiness Zones	3 to 8
Native Habitat and Range	Rich, deciduous or mixed coniferous woods and woodland coves in acid soils from Nova Scotia and Minnesota, south to Georgia and Missouri

Round-Lobed Hepatica

DESCRIPTION

Round-lobed hepatica has a rosette of basal, evergreen leaves with three round lobes. The furry, 3/4- to 1-inch flowers appear in early spring; they have 5 to 12 petallike sepals in shades of blue, white, or (rarely) pink. The foliage turns bronzy in the autumn.

GARDEN USES

Plant round-lobed hepatica on a mound or against a rock where the early flowers are sure to be noticed. Combine it with evergreen wild gingers (*Asarum* spp.), bloodroot (*Sanguinaria canadensis*), trilliums, rue anemone (*Anemonella thalictroides*), and Allegheny spurge (*Pachysandra procumbens*). The flowers look lovely with early spring bulbs such as Dutchman's breeches (*Dicentra cucullaria*), spring beauties (*Claytonia* spp.), species crocus, snowdrops, puschkinia, and corydalis.

GROWING AND PROPAGATION

Plant in humus-rich, moist, acid soil in light to full shade. Plants tolerate deep shade because of their evergreen leaves. They establish slowly but in time form multicrowned clumps. To propagate, lift plants after the new leaves are fully formed or in the fall, and gently pull the crowns apart. Sow fresh seed outdoors uncovered. Seedlings take 2 to 3 years to bloom.

RECOMMENDED CULTIVARS OF ROUND-LOBED HEPATICA

'Eco Blue Harlequin' has blue flowers and striking, silver-mottled leaves.

'Eco Indigo' has the deepest blue flowers available.

'Eco Tetra Blue' has 1½-inch, deep blue flowers.

ANOTHER *HEPATICA* SPECIES

Sharp-lobed hepatica (*Hepatica acutiloba*) has larger leaves than *H. americana*, up to 5 inches, with pointed lobes. Flower color varies from white, pink, or blue to lavender or purple. Multipetaled and double-flowered forms are available. 'Eco Regal Blue' has rich blue flowers more than 1 inch wide. 'Eco White Fluff' has large, shaggy, double flowers. 'Eco White Giant' has sparkling white flowers 1¾ inches across and extra lobes on the leaves. Rich, moist, limy soil in light to full shade. Found in rich, deciduous woods and woodland coves in limy soil from Quebec and Minnesota, south to Georgia and Missouri. Zones 3 to 8.

Heuchera americana

American Alumroot, Rock Geranium

Pronunciation	HUE-ker-uh uh-mare-ih-KAH-nuh
Family	Saxifragaceae, Saxifrage Family
USDA Hardiness Zones	4 to 9
Native Habitat and Range	Open woods, rocky slopes, outcroppings, and embankments from New England and Michigan, south to Georgia and Oklahoma

American Alumroot

DESCRIPTION

American alumroot is an open, mounding plant with mottled, silvery green, heart-shaped leaves with scalloped edges. The leaves have long, slender leafstalks that arise directly from a thick, woody crown. Airy sprays of tiny green flowers open on 1½- to 3-foot naked stalks above the leaves in spring. In the fall, the leaves turn ruby-red or purplish.

GARDEN USES

To make the most of the beautiful foliage, plant alumroot at the front of beds and borders with low to medium-size plants such as moss phlox (*Phlox subulata*), sedums, hardy cranesbills (*Geranium* spp.), creeping baby's-breath (*Gypsophila repens*), pinks (*Dianthus* spp.), and Persian catmint (*Nepeta mussinii* 'Blue Wonder'). In shaded spots plant it with ferns, sedges (*Carex* spp.), creeping phlox (*Phlox stolonifera*), bleeding hearts (*Dicentra* spp.), and irises.

GROWING AND PROPAGATION

Plant alumroots in moist but well-drained, humus-rich soil in full sun or partial shade. Leaves will bleach if exposed to hot afternoon sun. Divide and replant older clumps every few years because the crowns tend to rise above the soil as they age. Grow alumroot from seeds sown inside or outside. The seeds are small, so cover them lightly. Seedlings develop slowly at first. Cultivars will not come true from seed. Borers may burrow into the crowns of old alumroots and kill them. If plants are declining, check for borer damage, and dig and destroy infested plants.

RECOMMENDED CULTIVARS OF AMERICAN ALUMROOT

'Dale's Strain' is a seed-grown strain with gray-green leaves with silver mottling.

There are several excellent hybrids between 'Dale's Strain' and other heucheras, including red-leaf forms such as 'Garnet', 'Montrose Ruby', 'Ruby Veil', and 'Chocolate Ruffles'. Silver-leaf selections include 'Persian Carpet' and 'Pewter Veil'.

ANOTHER *HEUCHERA* SPECIES FOR FOLIAGE INTEREST

Hairy alumroot (*Heuchera villosa*) has 8-inch, maplelike leaves that are hairy but soft. The small, greenish white flowers appear in 1- to 2-foot, airy clusters. A form with purple foliage is available. Plant in humus-rich, moist but well-drained soil in sun or shade. Found in rocky woods and on cliffs and outcroppings from Virginia and Indiana, south to South Carolina and Alabama. Zones 4 to 8.

Hibiscus moscheutos

Rose Mallow, Marsh Mallow

Pronunciation	hy-BISS-kus moss-SHOO-toes
Family	Malvaceae, Mallow Family
USDA Hardiness Zones	Zones 4 (with protection) or 5 to 10
Native Habitat and Range	Wet meadows, low woods, marshes, and ditches from Maryland and Ohio, south to Indiana and Texas

Rose Mallow

DESCRIPTION

Rose mallow is a shrublike perennial with broad, oval leaves that have three to five shallow lobes. Its erect stalks arise from a woody crown with deep, spreading roots. The open clusters of 6- to 8-inch white flowers with deep red centers appear at the tops of the stems for several months in summer. Individual flowers last only one day, but open in succession. The woody seed capsules add fall and winter interest.

GARDEN USES

Rose mallow is a great accent plant—it livens up borders when combined with ornamental grasses and airy summer perennials. Combine it with Culver's root (*Veronicastrum virginicum*), bee balms (*Monarda* spp.), Kansas gayfeather (*Liatris pycnostachya*), flowering spurge (*Euphorbia corollata*), and ironweeds (*Vernonia* spp.). Or plant it as a shrub surrounded by fine-textured flowers and grasses. In meadow or waterside gardens, combine rose mallow with irises, ferns, astilbes, and ornamental grasses.

GROWING AND PROPAGATION

Plant rose mallow in evenly moist, humus-rich soil in full sun or light shade. Set out young plants in spring or fall. Leave at least 3 feet between plants to allow for their eventual spread. Established plants do not transplant well. If you must divide them, cut the tough, woody clumps apart. Propagate by tip cuttings taken in early summer. Reduce the leaf size on the cuttings by cutting individual leaves in half, unless you can keep the cuttings under mist. Cultivars do not come true from seed, but most seedlings will produce attractive plants. Sow seeds outdoors in fall.

Japanese beetles love rose mallow. They feed on leaf tissue, leaving behind a lacelike leaf skeleton. To control them, pick the beetles off by hand and drop them into a pail of soapy water.

RECOMMENDED HYBRID MALLOW CULTIVARS

Some hybrid mallow cultivars have 8- to 10-inch flowers in a variety of colors. **'Sweet Caroline'** has ruffled pink flowers with a dark eye. **'Anne Arundel'** has 9-inch, pink flowers. **'Lady Baltimore'** is pink with a deep red center. **'Lord Baltimore'** is deep scarlet.

Iris cristata

Crested Iris

Pronunciation	EYE-ris kris-TAH-tuh
Family	Iridicaeae, Iris Family
USDA Hardiness Zones	3 to 9
Native Habitat and Range	Rich, deciduous woods, rocky wooded slopes, outcroppings, and shady roadsides in acid soils from Maryland and Ohio, south to Georgia and Oklahoma

Crested Iris

DESCRIPTION

Crested iris is a small, spring-blooming iris for woodland and rock gardens. This creeping iris grows from thick, fleshy rhizomes and has fans of short, broad leaves. The foliage stands only 4 to 8 inches tall when mature. The flattened, sky blue flowers are 2 inches wide and have a yellow and white blaze.

GARDEN USES

Plant crested iris where it can form broad mats of foliage, such as in rock gardens or at the front of a bed along a path. Or, use it as a groundcover under shrubs and flowering trees. Try combining crested iris with twinleaf (*Jeffersonia diphylla*), ferns, spring beauties (*Claytonia* spp.), wild gingers (*Asarum* spp.), creeping phlox (*Phlox stolonifera*), bloodroot (*Sanguinaria canadensis*), wild cranesbill (*Geranium maculatum*), hepatica (*Hepatica* spp.), creeping Jacob's ladder (*Polemonium reptans*), and merrybells (*Uvularia* spp).

GROWING AND PROPAGATION

Plant in rich, moist, slightly acid soil in light to partial shade. Crested iris blooms best when it receives some direct summer sun. Plants grow to form extensive mats. Divide plants in late summer; replant divisions with the top of the rhizome just above the soil surface. Sow fresh seeds outdoors.

Bacterial rot may infect irises, causing brown and black spots on the leaves. To prevent bacterial rot, don't bury rhizomes too deep and don't plant in wet soil. Remove and destroy infected leaves.

RECOMMENDED SELECTIONS OF CRESTED IRIS

'Abbey's Violet' has deep blue-violet flowers.

'Eco Little Bluebird' is a dwarf that grows to 4 inches tall with deep blue flowers.

I. cristata* var. *alba (sold as 'Alba') is pure white, but many white-flowered forms bear this name.

'Shenandoah Sky' has rich sky blue flowers.

'Summer Storm' has deep blue flowers.

ANOTHER *IRIS* SPECIES FOR WOODLAND AND ROCK GARDENS

Vernal iris or **violet iris** (*Iris verna*) bears 2-inch, blue-violet flowers on 4-inch stalks in early spring. The grasslike leaves elongate after flowering to 1 foot. Plant in average, sandy or loamy, well-drained soil in full sun or partial shade. Open, gravelly slopes, open woods, and pinelands from Pennsylvania and Kentucky, south to Georgia and Mississippi. Zones 4 to 8.

Iris versicolor

Blue Flag Iris

Pronunciation	EYE-ris VER-suh-kuh-ler
Family	Iridicaeae, Iris Family
USDA Hardiness Zones	2 to 8
Native Habitat and Range	Wet meadows and prairies, still ponds, shallow marshes, and bogs from Newfoundland and Manitoba, south to Virginia and Minnesota

Blue Flag Iris

DESCRIPTION

Blue flag iris is a thick, leafy iris that thrives in wet conditions. Plants have bold, straplike leaves to 3 feet long. In early summer, bright blue-violet flowers open atop 1½- to 3-foot flowerstalks. There are many color forms ranging from white to sky blue and purple. 'Kermesina' is a seed-grown strain with red-violet flowers.

GARDEN USES

Choose blue flag iris for pondside plantings with ferns, grasses, and lush flowering perennials. It combines beautifully with plants that have rounded forms and bold flowers, including hibiscus, bee balm (*Monarda didyma*), swamp milkweed (*Asclepias incarnata*), and ironweeds (*Vernonia* spp). In water gardens, plant blue flag iris with wild calla (*Calla palustris*), pickerel weed (*Pontederia cordata*), arrowheads (*Sagittaria* spp.), and marsh marigold (*Caltha palustris*). It also grows well in beds and borders in moist soil.

GROWING AND PROPAGATION

Plant in evenly moist to wet, humus-rich soil in full sun or light shade. To grow blue flag iris in a water garden, plant in containers containing rich, clayey soil, cover the soil with 2 inches of pea gravel, and submerge it in up to 8 inches of water. Propagate by dividing plants after flowering in summer or early fall. Replant rhizomes immediately in well-prepared soil. Sow ripe seeds outdoors. Seeds will germinate the following spring.

OTHER *IRIS* SPECIES FOR WET CONDITIONS

Copper iris (*Iris fulva*) is a unique, red-flowered species that blooms in early summer. It grows 3 to 4 feet tall, with flowers to 3½ inches wide. The flowers have small standards and drooping falls, so they appear flat. Plant in moist to wet, humus-rich soil in full sun or light shade. Wet meadows and swamps from Illinois and Missouri, south to Georgia and Texas. Zones 4 (with winter protection) or 5 to 10.

Missouri iris or **Rocky mountain iris** (*Iris missouriensis*) is a delicate species with narrow, straplike leaves 1 to 1½ feet long. The white to deep blue summer flowers have slender standards and falls borne above the leaves on 2-foot stems. Plant in rich, moist soil in full sun or light shade; it will tolerate dry sites. Wet meadows, streamsides, and seasonally wet areas from South Dakota and British Columbia, south to Mexico and California. Zones 3 to 8.

Southern blue flag or **Virginia iris** (*Iris virginica*) resembles blue flag iris, but the flowers are carried even with the 2- to 3-foot leaves. The leaf tips droop downward. Plant in evenly moist to wet, humus-rich soil in full sun or light shade. Low meadows and wetlands on the coastal plain from Virginia to Texas. Zones 4 to 9.

Isopyrum biternatum

False Rue Anemone

Pronunciation	eye-sow-PIE-rum bye-ter-NAY-tum
Family	Ranunculaceae, Buttercup Family
USDA Hardiness Zones	3 to 8
Native Habitat and Range	Rich, deciduous woods, woodland coves, floodplains, and streamsides in limy soils from Ontario and Minnesota, south to Florida and Arkansas

False Rue Anemone

DESCRIPTION

False rue anemone has delicate basal and stem leaves with small, round-lobed leaflets. The leaves are divided two times into six clusters with three leaflets each. Sparse clusters of snow white, ¾-inch flowers cover the plants for a month in spring. Each flower consist of five petal-like sepals. Plants grow from creeping roots with thickened portions that form new crowns. Its name is appropriate because its foliage and flowers resemble those of rue anemone (*Anemonella thalictroides*).

GARDEN USES

False rue anemone in flower can turn a woodland as white as a late-spring snow. The plants spread through and around other species but are seldom a nuisance. Combine them with shade-loving wildflowers and perennials such as wild ginger (*Asarum* spp.), phlox, bloodroot (*Sanguinaria canadensis*), wild cranesbill (*Geranium maculatum*), irises, primroses, epimediums (*Epimedium* spp.), and lungworts (*Pulmonaria* spp).

GROWING AND PROPAGATION

Plant in humus-rich, evenly moist soil in light to full shade. Plants disappear after flowering but new foliage may emerge in fall or late winter. Large colonies form from creeping roots and self-sown seedlings. Divide plants after flowering. Sow fresh seeds outdoors.

PERFECT GARDEN COMPANIONS

Dwarf larkspur (*Delphinium tricorne*) is a succulent, spring-blooming delphinium with 1- to 1½-inch, deep blue, spurred flowers and three- to five-lobed leaves on 1- to 3-foot stalks. Plants go dormant after flowering. Dwarf larkspur is a plant of sunny spring woodlands that become shaded in summer. Plant them in clumps so they will show up in the broken light of dappled shade. Combine them with false rue anemone and other plants that need limy soil such as trilliums, merrybells (*Uvularia* spp.), spring beauties (*Claytonia* spp.), wild gingers (*Asarum* spp.), and ferns. Plant in rich, moist, limy soils in full sun to partial shade. They increase very slowly from branches on the crown and seldom need division. Found in rich, moist, deciduous woods and bottomlands in limy soils. Zones 4 to 7.

Zigzag goldenrod (*Solidago flexicaulis*) has thin, wiry 1- to 3-foot stems that bend back and forth between the nodes of alternate, rounded leaves, giving the stem a crooked, zigzag appearance. The starry yellow flowers intermingle with the smaller stem leaves along the upper third of the stem. The blooms add late-season color to shady spots, and in summer the foliage fills the gap left when false rue anemone goes dormant. Plant in rich, moist soil in light to full shade. Like all goldenrods, they are long-lived garden plants that thrive on neglect. Plants form multistemmed clumps that are never invasive. Found in rich, deciduous woods and clearings. Zones 3 to 8.

Jeffersonia diphylla

Twinleaf

Pronunciation	jef-fer-SOH-nee-uh die-FILL-uh
Family	Berberidaceae, Barberry Family
USDA Hardiness Zones	4 to 8
Native Habitat and Range	Rich, deciduous woods, woodland coves, and streamsides in limy soils from Ontario and Minnesota, south to Virginia and Alabama

Twinleaf Flowers

DESCRIPTION

Twinleaf gets its name from its paired leaflets on slender stalks. The 6-inch, sea green leaves arise like forest butterflies from a basal clump, growing up to 1½ feet tall. The naked flowerstalks bear single, pure white, eight-petaled flowers in early spring. Twinleaf flowers die off quickly, followed by pipelike seed capsules in early summer. Plants produce tangled wiry roots from dense crowns.

Twinleaf Seed Capsule

GARDEN USES

Twinleaf's sturdy, handsome foliage looks beautiful in shady spots with ferns, spring beauties (*Claytonia* spp.), wild ginger (*Asarum* spp.), phlox, bloodroot (*Sanguinaria canadensis*), wild cranesbill (*Geranium maculatum*), hepatica (*Hepatica* spp.), creeping Jacob's ladder (*Polemonium reptans*), merrybells (*Uvularia* spp.), and irises. The early spring flowers open with ephemeral species such as spring beauties, toothworts (*Dentaria* spp.), dutchman's breeches (*Dicentra cucullaria*), and trout lilies (*Erythronium* spp.).

GROWING AND PROPAGATION

Plant in humus-rich, moist, limy soil in partial to full shade. Plants grow slowly but form broad clumps over time. Divide the crowns in fall using a sharp knife to cut the tough crown. Leave two or three new buds per clump. Sow fresh seeds outdoors. Plants bloom in 2 to 3 years.

Liatris aspera

Rough Gayfeather, Button Gayfeather

Pronunciation	lee-AH-tris AS-per-uh
Family	Asteraceae, Aster Family
USDA Hardiness Zones	3 to 9
Native Habitat and Range	Dry prairies, savannas, and plains from Ontario and North Dakota, south to South Carolina and Texas

Rough Gayfeather

DESCRIPTION

Rough gayfeather bears clusters of 1-inch, buttonlike flowers on short stalks at the top of 4- to 6-foot flower spikes. The pale purple or pink flowers open in mid- to late summer. Its gray-green, grasslike leaves, which grow up to 16 inches long, twist in a clockwise circle as they leave the stem. Plants sprout from fat underground stems called corms.

GARDEN USES

Rough gayfeather is a magnet for monarch butterflies. They cover the tall flowerstalks by the dozens in summer. Choose rough gayfeather for informal meadow and prairie gardens as well as formal borders. Combine it with asters, goldenrods, butterfly weed (*Asclepias tuberosa*), prairie clovers (*Dalea* spp.), and purple coneflowers (*Echinacea* spp.). In borders, plant rough gayfeather with garden phlox (*Phlox paniculata*), yarrows, artemisias (*Artemisia* spp.), Shasta daisies (*Chrysanthemum mathx superbum*), and ornamental grasses.

GROWING AND PROPAGATION

Plant in sandy or loamy, moist but well-drained soil in full sun. Rough gayfeather occurs in dry, sandy soils in the wild and may overgrow and flop in rich, moist soil. Sow ripe seeds outdoors. Germination will occur the next spring. Indoors, stratify the seeds at 40°F for 4 to 6 weeks to encourage even germination. Plants bloom in 2 to 4 years.

OTHER *LIATRIS* SPECIES WITH BUTTONLIKE FLOWERS

Cylindric blazing-star (*Liatris cylindracea*) grows from 8 to 24 inches tall with narrow clusters of pale purple flowers in open spikes. Its leaves grow to 10 inches long. It blooms in late summer. Plant in average, sandy, or loamy, well-drained soil in full sun or light shade. Dry prairies, savannas, and open woods. Zones 3 to 9.

Rocky Mountain blazing-star (*Liatris ligulistylis*) is similar to cylindric blazing-star, with dark violet flowers that appear broader and more open. Plants grow 2 to 5 feet tall. Plant in humus-rich, evenly moist soil in full sun. Wet, black soil prairies and borders of marshes. Zones 3 to 8.

Scaly blazing-star (*Liatris squarrosa*) has round, violet flowers borne in open clusters on 1- to 3-foot stalks. The stiff leaves are sparse, and they mix with the flowers to the top of the stalk. Plant in average, well-drained soil in sun or light shade. Open woods, meadows, and roadsides. Zones 4 to 9.

Liatris pycnostachya

Kansas Gayfeather

Pronunciation	lee-AH-tris pick-no-STAKE-ee-uh
Family	Asteraceae, Aster Family
USDA Hardiness Zones	3 to 9
Native Habitat and Range	Moist, black soil prairies, low meadows, and moist savannas from Indiana and South Dakota, south to Louisiana and Texas

Kansas Gayfeather

DESCRIPTION

Kansas gayfeather is a giant wildflower, with 3- to 5-foot spikes of densely packed, red-violet to mauve flowers on stiff, leafy stems. The deep green basal leaves are up to 12 inches long. Leaves further up the stems are smaller, blending into the flowers. Plants bloom in midsummer. The variety *alba*, also sold as 'Alba', has creamy white flowers, but it is produced from seeds and flower color is variable. Plants sprout from fat, underground stems called corms.

GARDEN USES

Plant Kansas gayfeather in meadow and prairie gardens with mountain mints (*Pycnanthemum* spp.), purple coneflowers (*Echinacea* spp.), wild bergamot (*Monarda fistulosa*), goldenrods, milkweeds (*Asclepias* spp.), rosinweed (*Silphium integrifolium*), gray-headed coneflower (*Ratibida pinnata*), and ornamental grasses. You can also use them in beds and borders with a variety of summer-blooming perennials.

GROWING AND PROPAGATION

Plant in rich, evenly moist soil in full sun. They often need support from other plants or staking to keep the stems erect. Give them plenty of room without competition from other plants to keep them vigorous. Clumps seldom need division to maintain plant vigor, but you can propagate new plants by dividing the corms in early fall. You can also grow Kansas gayfeather from seed, with the same pretreatment and sowing as described for *L. aspera*.

Mice and voles love to eat the corms of Kansas gayfeather. To prevent damage, dig out an area and line it with hardware cloth prior to planting the corms.

ANOTHER *LIATRIS* SPECIES WITH SPIKY FLOWERS

Spike gayfeather (*Liatris spicata*) has compact, 2- to 4-foot stems with terminal spikes of deep red-violet flowers in narrow, tightly packed heads. The leafy stems are stiff and seldom need support. Excellent cultivars are available in a variety of colors and sizes. 'August Kobold', the most popular cultivar, has dense spikes of mauve to violet flowers on 2- to 2½-foot stems. Plant in average to rich, well-drained soil in full sun. Moist meadows and prairies and rocky mountainsides from New York and Michigan, south to Florida and Louisiana. Zones 3 to 8.

Lilium superbum

Turk's-Cap Lily

Pronunciation	LILL-ee-um soo-PER-bum
Family	Liliaceae, Lily Family
USDA Hardiness Zones	4 to 9
Native Habitat and Range	Open, deciduous woods, meadows, clearings, and roadsides from New Brunswick to Indiana, south to Florida and Missouri

DESCRIPTION

Turk's-cap lily is a slender plant that can reach 7 feet. It has tiered whorls of broad, 6-inch lancelike leaves and large clusters of spotted, bright orange flowers. The flower petals curve downward strongly, and each flower has a distinct green star in the throat.

GARDEN USES

Lilies add elegance to the garden. Use Turk's-cap lily to accent shrubs and small, flowering trees. Plant it in open shade with irises, ferns, hostas, and other foliage plants. In woodland gardens, combine it with bee balm (*Monarda didyma*), ragged coneflower (*Rudbeckia laciniata*), Joe-Pye weeds (*Eupatorium* spp.), tall bellflower (*Campanula americana*), asters, goldenrods, turtleheads (*Chelone* spp.), and grasses.

GROWING AND PROPAGATION

Plant in humus-rich, evenly moist but well-drained, acid soil in full sun or partial shade. Plants bloom better with some direct sun. Good drainage is essential to avoid rot, especially in winter. Stake plants to prevent them from toppling over or breaking off near the base. Insert the stake near the stem, but take care not to spear the bulb with the stake. Attach the stems loosely to the stake. Propagate by dividing bulbs in late summer as they go dormant. Replant immediately. Sow ripe seed outdoors. Seeds need a combination of cool and warm treatments to germinate, so seedlings may take two seasons to emerge. Plants take 5 to 7 years to bloom from seed.

Turk's-Cap Lily

Lilies are susceptible to fungal and bacterial infections that damage or destroy the bulb. Providing proper growing conditions is the best way to prevent problems. If leaves develop yellow mottling, chances are the plants have a viral infection, which is not treatable. Dig and destroy all infected plants. Rodents may eat lily bulbs. To protect the bulbs, sprinkle a repellent around the area after planting.

OTHER LARGE *LILIUM* SPECIES

Canada lily (*Lilium canadense*) is a tall, slender lily that grows to 5 feet with whorls of narrow leaves. The nodding, bell-shaped flowers cluster at the top of the stems. Flowers may be yellow, red, or orange. Plant in humus-rich, evenly moist soil in full sun or light shade. Wet meadows, open woods, and ditches from Maine and Wisconsin, south to Alabama and Indiana. Zones 3 to 7.

Gray's lily or **bell lily** (*Lilium grayi*) is a striking plant with 3- to 4-foot stems bearing distinct whorls of broad, lancelike, dark green leaves. The 2-inch, bell-like flowers are deep crimson and face outward. They are carried in open clusters. Plant in rich, moist, but well-drained soil in full sun to partial shade. Meadows, clearings, open woods, and roadsides in the mountains of Virginia, North Carolina, and Tennessee. Zones 5 to 8.

Linnaea borealis

Twinflower

Pronunciation	lin-NAY-ee-uh bore-ee-AL-is
Family	Caprifoliaceae, Honeysuckle Family
USDA Hardiness Zones	1 to 8
Native Habitat and Range	Cool, moist, mixed deciduous or coniferous forests; also bogs and clearings from near the North Pole, south to West Virginia, Minnesota, and California

Twinflower

DESCRIPTION

Twinflower has tiny, rounded leaves with shallow teeth at the tips. The leaves occur in pairs along the trailing stems. The ¾-inch, rosy-pink flowers nod opposite one another on threadlike stems. The stems root at the nodes as they spread.

GARDEN USES

Twinflower forms a delicate, low groundcover under shrubs and in shaded gardens with rich, acid soil. Plant it in an old stump or trailing along a rotted log. Combine twinflower with wintergreen (*Gaultheria procumbens*), Canada mayflower (*Maianthemum canadense*), bunchberry (*Cornus canadensis*), partridgeberry (*Mitchella repens*), goldthread (*Coptis groenlandica*), round-lobed hepatica (*Hepatica americana*), and woodland asters.

GROWING AND PROPAGATION

Plant in humus-rich, moist, acid soil in sun or shade. Plants form dense colonies loaded with flowers, especially when they receive some direct sun. Divide clumps in spring or fall. Take stem cuttings in early summer. Sow seeds outdoors in fall or indoors with 4 weeks of cold, moist stratification.

PERFECT GARDEN COMPANIONS

Bead lily, bluebead (*Clintonia borealis)* is a lush foliage plant with naked stalks topped by four- to eight-flowered clusters of ¾-inch, nodding chartreuse bells. The deep, glossy blue berries form in late summer. Plant them with other northern wildflowers such as twinflower, which makes an dainty carpet under the bold leaves. Starflower (*Trientalis borealis*), sedges (*Carex* spp.), and ferns are also good companions. Plant them in a rotted pine stump to raise the foliage and flowers for added interest. Bead lilies are not for every garden. They demand cool, rich, acid soil in partial to full shade. Plants will perish in less than optimal conditions. Found in rich, acidic boreal and mixed coniferous woods and bogs. Zones 2 to 7.

Speckled wood lily (*Clintonia umbellulata*) has similar foliage to bead lily but carries its 10 to 30, upward-facing, flat, open, white flowers in dense clusters at the top of 10- to 12-inch stems. The black berries ripen in late summer and persist into the winter if not eaten by birds. Plant in moist, well-drained, humus-rich, acid soil in partial to full shade. Found in rich woods and woodland coves and on mountain slopes. Zones 5 to 7.

Lobelia cardinalis

Cardinal Flower

Pronunciation	low-BEE-lee-uh car-dih-NAH-lis
Family	Campanulaceae, Bellflower Family
USDA Hardiness Zones	2 to 9
Native Habitat and Range	Low meadows, open, wet woods, swamps, and open sandbars in streams and rivers from New Brunswick and Minnesota, south to Florida and the Gulf states

Cardinal Flower

DESCRIPTION

The deep blood red flowers of cardinal flower add color to moist meadows, water gardens, and borders in summer. The plants form rosettes in fall that remain green throughout the winter except in very cold climates. In spring, the leafy flowerstalks arise from the crown. The dense, erect flower spikes have irregular, tubular flowers for 2 to 3 weeks in mid- to late summer. The flowers fade to form button-shaped seed capsules. 'Heather Pink' has soft pink flowers and comes true from seed. 'Royal Robe' has ruby red flowers on sturdy, 3-foot stems.

GARDEN USES

Combine cardinal flower with blue flag iris (*Iris versicolor*), sneezeweeds (*Helenium* spp.), monkey flowers (*Mimulus* spp.), ironweeds (*Vernonia* spp.), and sedges (*Carex* spp.). Plant it by water gardens with astilbes, ferns, and bold foliage plants such as bigleaf ligularia (*Ligularia dentata*) and hostas. Cardinal flower will grow well in borders as long as the soil remains moist. Combine it with cannas (*Canna* spp.), hibiscus, daylilies, turtleheads (*Chelone* spp.), garden phlox (*Phlox paniculata*), and ornamental grasses.

GROWING AND PROPAGATION

Plant in rich, constantly moist soil in full sun to partial shade. Plants in warmer zones with fluctuating winter temperatures often rot if mulched, but in colder zones a winter mulch is mandatory. Remove the mulch as soon as the weather warms to avoid killing the rosettes. Plants are often short-lived; dividing them frequently keeps them vigorous. Lift clumps in early fall and remove the new rosettes from the old rootstocks. Replant immediately in enriched soil. Plants self-sow prolifically where the soil is bare in fall and winter. Sow seeds uncovered outdoors or indoors in early spring. They germinate quickly and bloom the first season.

OTHER *LOBELIA* SPECIES

Great blue lobelia (*Lobelia siphilitica*) has deep blue flowers up to 1 inch long in midsummer. The plants grow 2 to 3 feet tall and the leaves to 5 inches long. Plant in rich, moist soil in full sun or light shade. Low meadows, bottomlands, and riverbanks. Zones 3 to 8.

Spiked lobelia (*Lobelia spicata*) bears ¼-inch, pale blue to white flowers on leafless spikes. Plants grow 1 to 2½ feet tall. Plant in rich, moist soil in full sun or light shade. Moist prairies, low meadows, clearings, and marsh edges. Zones 3 to 8.

Lupinus perennis

Wild Lupine

Pronunciation	lew-PIE-nus per-EN-iss
Family	Fabaceae, Pea Family
USDA Hardiness Zones	3 to 9
Native Habitat and Range	Open pine and oak woods, dry prairies and meadows, savannas, and clearings in sterile, acid soil from Maine and Minnesota, south to Florida and Louisiana

Wild Lupine

DESCRIPTION

Wild lupine is a delicate creeping plant with narrow, blunt leaflets that form a circular fan on 3- to 4-inch leaf stems. The rosettes of leaves arise from crowns at the end of the creeping stems. The medium blue flowers, which resemble sweet pea blossoms, bloom on open, 12-inch flower spikes in late spring and early summer.

GARDEN USES

Plant wild lupine in wildflower meadows and prairie gardens, on sandy banks, or along roadsides. Combine it with downy phlox (*Phlox pilosa*), spiderworts (*Tradescantia* spp.), puccoons (*Lithospermum* spp.), gayfeathers (*Liatris* spp.), and grasses. For borders and flowerbeds, choose hybrid lupines rather than wild lupine.

GROWING AND PROPAGATION

Wild lupine is difficult to establish. Plant in well-drained, acid, sandy soil in sun or light shade. For best results, start plants from seed. Inoculate seed with a bacterium specified for lupines (usually available from your seed supplier), and give seeds cold, moist stratification for 4 to 6 weeks. The seedlings will bloom in 2 years. Plants may be short-lived when planted in sites that don't suit their needs. Don't transplant or divide plants once they are established.

OTHER *LUPINUS* SPECIES

Texas bluebonnet (*Lupinus texensis*) is an annual that produces dense spring clusters of deep blue flowers on bloom stalks to 16 inches tall. The foliage is similar to wild lupine but smaller. Plant in average to rich, well-drained soil in full sun or light shade. Sow seed outdoors where plants are to grow. Bluebonnets reseed readily in favorable conditions. Found in open woods, fields, and roadsides in eastern and central Texas. Zones 6 to 9.

Washington lupine or **garden lupine** (*Lupinus polyphyllus*) grows 3 to 5 feet tall with dense flower spikes of white, pink, or blue flowers. The long, pointed leaflets fan out at the end of 6- to 8-inch leaf stems. Plant in rich, moist, acid soil in full sun or light shade. Native to the west coast of North America but widely naturalized. Zones 3 to 7.

Maianthemum canadense

Canada Mayflower

Pronunciation	my-AN-thuh-mum can-uh-DEN-see
Family	Liliaceae, Lily Family
USDA Hardiness Zones	2 to 8
Native Habitat and Range	Moist, deciduous or coniferous woods, lake shores, and clearings in acid soil from Newfoundland and the Northwest Territories, south to New Jersey and North Dakota, and in the mountains to North Carolina

Canada Mayflower

DESCRIPTION

Canada mayflower forms an open groundcover of glossy, green foliage in cool, wooded sites. The 2- to 4-inch basal leaves are heart-shaped; the 2- to 3-inch stem leaves are narrower. Dense spikes of tiny, white flowers with protruding, fuzzy stamens blanket the plants in midsummer. Flowers fade to reveal green-speckled, red berries that attract wildlife. Plants spread by creeping underground rhizomes.

GARDEN USES

Canada mayflower weaves beautifully through other woodland plants in shade gardens with acid soil. Combine it with galax (*Galax rotundifolia*), wintergreen (*Gaultheria procumbens*), vernal iris (*Iris verna*), bunchberry (*Cornus canadensis*), twinflower (*Linnaea borealis*), round-lobed hepatica (*Hepatica americana*), and wild gingers (*Asarum* spp.).

GROWING AND PROPAGATION

Plant in humus-rich, moist, acid soil in sun or shade. Plants often display far more foliage than flowers, especially in dense shade. Divide clumps in fall. Remove pulp from the seeds and sow seeds outdoors in fall. Plants take several years to bloom.

PERFECT GARDEN COMPANIONS

Dewdrop, false violet (*Dalibarda repens*) is a ground-hugging, evergreen creeper with hairy, heart-shaped leaves. In early summer, delicate, five-petaled white flowers are held singly on thin stalks above the leaves. Plants grow amid other creepers such as Canada mayflower in rich coniferous woods. Plant them in moist, humus-rich, acid soils in light to full shade. Zones 3 to 6.

Rosy twisted stalk (*Streptopus roseus*) has succulent stems clothed in clasping, deeply veined, oval to lance-shaped succulent leaves that have a satiny finish. The stem branches into two divisions below the middle, and in spring and summer the small rose-pink bells nod from the leaf axils along the two stems. Plant in humus-rich, evenly moist, acid soil in full shade. Found in deep, deciduous or mixed coniferous woods growing with Canada mayflower and other northern wildflowers. Zones 3 to 7.

Sweet white violet (*Viola blanda*) is a delicate plant with thin, pleated leaves and tiny, ¼-inch, sweet-scented white flowers. Plants grow only 2 to 3 inches tall and spread by runners to form dainty colonies. Plant in humus-rich, moist to wet, acid soil in light to full shade. Found in wet woods, ravines, seeps, and bogs. Reclassified as *Viola macloskeyi* ssp. *pallens* but seldom listed that way. Zones 3 to 8.

Mertensia virginica

Virginia Bluebells

Pronunciation	mer-TEN-see-uh ver-JIN-ih-kuh
Family	Boraginaceae, Borage Family
USDA Hardiness Zones	3 to 9
Native Habitat and Range	Moist, deciduous woods, floodplains, streamsides, and clearings from New York and Minnesota, south to Alabama and Kansas

Virginia Bluebells

DESCRIPTION

Virginia bluebells are spring bloomers that go dormant after blooming. They have a rosette of delicate, blue-green leaves and succulent stems that bear sky blue, bell-shaped flowers up to 1 inch long. The thick stems grow 1 to 2 feet, sprouting from stout, thickened roots. Leaves reach 8 inches long. Flower buds are pink before opening. The stems die down as soon as flowering is complete. 'Alba' has white flowers. Pink- and lavender-flowered selections are occasionally found in the wild and in gardens but are not commercially available.

GARDEN USES

Plant Virginia bluebells with other early blooming wildflowers, including shooting stars (*Dodecatheon* spp.), columbines (*Aquilegia* spp.), woodland phlox (*Phlox divaricata*), and trilliums. Virginia bluebells also look perfect in a shaded garden with spring bulbs such as daffodils and tulips. Pair them with plants that have persistent foliage, such as Canada wild ginger (*Asarum canadense*) and ferns, that will fill the gaps left when the Virginia bluebells go dormant (plant them beside, not on top of Virginia bluebells).

GROWING AND PROPAGATION

Plant Virginia bluebells in consistently moist but well-drained, humus-rich soil in sun or shade. Deep purple shoots emerge in early spring and quickly expand to reveal the buds. The flowers begin opening as the stems elongate. Flowering lasts several weeks, and the plants go dormant soon after. Take care not to dig into the dormant clumps accidentally. Propagate by dividing large clumps after flowering. Self-sown seedlings will appear. They bloom after 2 or 3 years.

ANOTHER *MERTENSIA* SPECIES

Tall lungwort (*Mertensia paniculata*) is an upright, succulent plant from 2 to 3 feet tall with puckered, oval, blue-green leaves with pointed tips. The ½-inch nodding flowers open in late spring and early summer. Plant in humus-rich, moist to wet, acid soil in full sun or partial shade. Plants grow naturally in mixed open woods, clearings, and areas around bogs from Hudson Bay, south to Michigan, Minnesota, and Washington. Zones 4 to 7.

Mitchella repens

Partridgeberry, Twinberry

Pronunciation	mit-CHELL-uh REH-penz
Family	Rubiaceae, Madder Family
USDA Hardiness Zones	3 to 9
Native Habitat and Range	Rich, deciduous or coniferous woods, pinelands, and clearings from Nova Scotia and Minnesota, south to Florida and Texas

Partridgeberry Flowers

Partridgeberry Berries

DESCRIPTION

Partridgeberry is a trailing plant with shiny, round or oval, evergreen leaves. Each leaf has a distinctive white stripe down its center vein. The paired, snow white, sweet-scented flowers have four fringed petals that form ½-inch stars in summer. The name twinberry arises from the fact that each flower pair forms one ruby-red fruit. The wintergreen-scented berries persist through the following summer after they form. The trailing stems root as they grow.

GARDEN USES

Choose partridgeberry as a groundcover under shrubs or to weave through plantings of other wildflowers, such as round-lobed hepatica (*Hepatica americana*), wintergreen (*Gaultheria procumbens*), Canada mayflower (*Maianthemum canadense*), merrybells (*Uvularia* spp.), sedges (*Carex* spp.), and wood ferns (*Dryopteris* spp.).

GROWING AND PROPAGATION

Plant in humus-rich, moist, acid soil in light to full shade. The stems root at the nodes as they trail over the ground. Plants will form broad clumps of matted foliage in time. Divide plants in spring or take stem cuttings in early summer. Remove pulp from the seeds and sow them outdoors in the fall.

A PERFECT GARDEN COMPANION

Goldthread (*Coptis groenlandicum*) is a diminutive forest dweller with trailing stems forming mats of 1-inch, three-lobed, evergreen leaves. Plants bloom in spring and early summer. The dainty, ½-inch, five-petaled, white flowers are like tiny jewels strewn across the forest floor. When goldthread is planted with partridgeberry, the two plants' similar flowers open at different times of the year, extending the seasonal interest of the garden. Plant in cool, moist, humus-rich, acid soil in partial to full shade. They are intolerant of summer heat and drought. Propagate by division in spring. Found in northern forests and mixed coniferous forests in acid soil. Zones 2 to 7.

Monarda didyma

Bee Balm

Pronunciation	mow-NAR-duh DID-ih-muh
Family	Lamiacea, Mint Family
USDA Hardiness Zones	3 to 8
Native Habitat and Range	Moist open woods, shaded roadsides, and clearings from Maine and Michigan, south to New Jersey and Ohio, and in the mountains to Georgia

Bee Balm

DESCRIPTION

Bee balm has 2- to 4-foot, succulent, square stems and opposite pairs of pointed, aromatic leaves 4 to 6 inches long. The 4-inch, spherical flower heads consist of tightly packed, brilliant scarlet blossoms surrounded by deep red modified leaves. Each tubular flower has a distinctive upper "lip" that arches out over the lower petals.

GARDEN USES

Bee balm is a beautiful summer perennial that attracts hummingbirds. Plant it in a moist, shaded spot or at the edge of a woodland with native lilies, Joe-Pye weeds (*Eupatorium* spp.), queen-of-the-prairie (*Filipendula rubra*), black-eyed Susans (*Rudbeckia* spp.), turtleheads (*Chelone* spp.), and ferns. In borders, combine bee balm with garden phlox (*Phlox paniculata*), yarrows, Shasta daisy (*Chrysanthemum* × *superbum*), cranesbills (*Geranium* spp.), and astilbes.

GROWING AND PROPAGATION

Plant in humus-rich, evenly moist soil in full sun or partial shade. Plants should not be allowed to dry out. Plants spread rapidly from creeping stems and die out in the center. Lift and divide entire clumps every 2 to 3 years to keep them vigorous. Replant vigorous portions in enriched soil.

Bee balm is susceptible to powdery mildew, which causes white blotches on the leaves. To minimize problems, buy mildew-resistant selections and keep plants evenly moist. Mildew resistance is often regional.

CULTIVARS OF BEE BALM

'Blue Stocking' has light violet-blue flowers with bright violet bracts and is mildew-resistant.

'Gardenview Scarlet' is mildew-resistant and has deep red flowers.

'Marshall's Delight' is a mildew-resistant selection with pure pink flowers.

'Snow White' has creamy white flowers on 3-foot stems.

ANOTHER *MONARDA* SPECIES

Wild bergamot (*Monarda fistulosa*) has lavender-pink to pale pink flowers with pink or white modified leaves, carried in 3-inch heads atop wiry 2- to 4-foot stems. The leaves are narrower, paler, and hairier than those of bee balm. 'Claire Grace' has lavender flowers and is mildew-resistant in the humid southeast. Plant in average to rich, well-drained soil in full sun to light shade. Plants grow naturally in meadows, prairies, old fields, clearings, and savannas and along roadsides from Quebec and British Columbia, south to Georgia and Arizona. Zones 3 to 9.

Oenothera fruticosa

Sundrops

Pronunciation	ee-no-THEE-ruh froo-tih-KOH-suh
Family	Onagraceae, Evening Primrose Family
USDA Hardiness Zones	4 to 8
Native Habitat and Range	Meadows, clearings, roadsides, and disturbed fields from Nova Scotia and New York, south to Florida and Alabama

DESCRIPTION

Sundrops have fragrant, bright yellow, 1½-inch flowers that open from red-tinged buds. The saucer-shaped flowers appear in clusters at the ends of the stems in mid- to late summer. The 2-foot stems may be upright or sprawling. They bear 1- to 3-inch, lancelike leaves covered with soft hairs. Rosettes of foliage remain green through the winter.

GARDEN USES

Sundrops add a dash of sunshine to meadows, rock gardens, and borders. Combine them with wild bergamot (*Monarda fistulosa*), garden phlox (*Phlox paniculata*), cranesbills (*Geranium* spp.), bellflowers (*Campanula* spp.), baptisias (*Baptisia* spp.), and gayfeathers (*Liatris* spp.). In tough spots plant them with ornamental grasses, yucca, barren strawberry (*Waldsteinia fragarioides*), and fire pinks (*Silene virginica*). They will bloom in a semishaded site with wild bleeding heart (*Dicentra eximia*), skullcaps (*Scutellaria* spp.), and ferns.

GROWING AND PROPAGATION

Plant in average to rich, well-drained soil in full sun or light shade. Plants tolerate partial shade in summer if they receive bright light in spring. Sundrops spread by slow creeping roots to form dense clumps. Plants spread rapidly in rich soil, and some may need to be dug out each season to keep them from overrunning neighboring plants. Divide the rosettes in early spring or after flowering in late summer.

Sundrops

ANOTHER FAST-GROWING *OENOTHERA* SPECIES

Common sundrops (*Oenothera tetragona*) is nearly identical to *O. fruticosa*, but the flower buds and stems are hairy, the leaves are broader, and the plants bear fewer flowers per stem. The foliage turns burgundy-red in fall. 'Fireworks' grows 18 inches tall and has reddish buds and stems. 'Sonnenwende' has white flowers that open in the evening on well-branched stems. 'Yellow River' grows to 2 feet tall and has bright yellow flowers. Plant in average to rich, well-drained soil in full sun or light shade. Plants grow naturally in open woods and clearings and along roadsides from Nova Scotia and Michigan, south to South Carolina and Louisiana. Zones 3 to 8.

Oenothera macrocarpa (also listed in the trade as *O. missouriensis*)

Ozark Sundrops, Missouri Primrose

Pronunciation	ee-no-THEE-ruh mack-row-CAR-puh
Family	Onagraceae, Evening Primrose Family
USDA Hardiness Zones	4 to 8
Native Habitat and Range	Open woods, cedar glades, savannas, and prairies in limy soil from Illinois and Colorado, south to Missouri and Texas

Ozark Sundrops

DESCRIPTION

Ozark sundrops are striking plants with large, 3- to 4-inch lemon yellow flowers. The flowers are saucer-shaped and have four wide, overlapping petals and a prominent stigma (female reproductive structure). Each flower lasts one day only. 'Greencourt Lemon' is more delicate than the species, with 2- to 2½-inch, soft, sulphur yellow flowers. The stems sprawl or grow weakly upright, reaching 6 to 12 inches tall. Lancelike, 2- to 4-inch, light green leaves cover the stems.

GARDEN USES

Ozark sundrops are lovely in both formal and wild settings. In prairies, meadows, and rock gardens plant them with penstemons (*Penstemon* spp.), gayfeathers (*Liatris* spp.), salvias, blackfoot daisy (*Melampodium leucanthum*), alumroots (*Heuchera* spp.), and grasses. In borders, combine them with cranesbills (*Geranium* spp.), yarrows, catmints (*Nepeta* spp.), purple coneflowers (*Echinacea* spp.), and irises.

GROWING AND PROPAGATION

Plant in average, sandy or loamy, well-drained soil in full sun or light shade. The plants are tough and drought-tolerant and require little care. Divide the multistemmed crowns in early spring or after flowering in late summer.

ANOTHER LOW-GROWING *OENOTHERA* SPECIES

Tufted evening primrose (*Oenothera caespitosa*) has spreading stems and grows to 8 inches tall. The lancelike, deep green leaves are 4 to 6 inches long and have toothed edges. Mature plants form dense, tufted clumps. The 2- to 3-inch flowers open white and turn pink by midday. Fresh flowers open in the late afternoon or evening. The faded flowers persist on the plant for at least a day before they fall off. Plant in average, rocky or sandy, well-drained soil in full sun. Dry banks, plains, and deserts in limy soils from Colorado and Washington, south to New Mexico and California. Zones 4 to 7.

Pachysandra procumbens

Allegheny Pachysandra

Pronunciation	pack-uh-SAN-druh pro-KUM-benz
Family	Buxaceae, Boxwood Family
USDA Hardiness Zones	4 to 9
Native Habitat and Range	Rich, deciduous or mixed coniferous woods and woodland coves in acid soil from North Carolina and Kentucky, south to Florida and Louisiana

DESCRIPTION

Allegheny pachysandra has succulent, leathery stems and broad, satiny, oval leaves that encircle the stems. The ragged flowers emerge and form dense spikes in the center of the leafy clumps in early spring. The new foliage emerges bright sea green with pale mottling. In the fall, frost brings out deep purple-blue background shades with silvery mottling. The plants form large clumps or extensive groundcovers from creeping stems.

GARDEN USES

Combine Allegheny pachysandra with flowering shrubs and wildflowers that favor acid soils, such as galax (*Galax rotundifolia*), shortia (*Shortia* spp.), partridgeberry (*Mitchella repens*), wild gingers (*Asarum* spp.), trilliums, and Allegheny foamflower (*Tiarella cordifolia*).

GROWING AND PROPAGATION

Plant in moist, humus-rich, acid soil in light to full shade. Plants spread slowly at first but form broad, rounded clumps in a few years. Divide clumps in spring before the new growth emerges. To take cuttings, don't cut stems with a knife. Instead, after new growth hardens in early summer, hold a stem at ground level and yank it firmly upward. You will get a bit of the old stem as well as some new roots. Treat the stem as a cutting until it is well rooted.

Allegheny Pachysandra

A CULTIVAR OF ALLEGHENY PACHYSANDRA

'Forest Green' has uniform, shiny green leaves in summer. This selection is also noteworthy because it is much easier to propagate than the species.

PERFECT GARDEN COMPANIONS

Nodding mandarin (*Disporum lanuginosum*) produces 1- to 3-foot, forked stems with rich green, oval leaves with pointed tips. In spring, one to three chartreuse flowers nod at the tips of the two branches. Nodding mandarin is showier in foliage than in flower. The plants make a nice vertical accent amid the low foliage of Allegheny pachysandra. Plant in evenly moist, humus-rich soil in partial to full shade. Plants eventually form multistemmed clumps that can be divided in autumn. Sow cleaned, fresh seed outdoors. Seeds have a complex dormancy and take one to two years to germinate. Found in rich, moist, deciduous or mixed coniferous woods. Zones 4 to 8.

Spotted mandarin (*Disporum maculatum*) is similar to nodding mandarin, but the flowers are creamy yellow spotted with purple. Plants grow 2 feet tall with multibranched, leafy stems. The red-orange berries are attractive in late summer as they hang at the tips of the stems. Plant in moist, humus-rich soil in shade. Zones 4 to 7.

Penstemon digitalis

Foxglove Penstemon

Pronunciation	PEN-steh-mon dij-ih-TAL-iss
Family	Scrophulariaceae, Figwort Family
USDA Hardiness Zones	3 to 8
Native Habitat and Range	Wet meadows and prairies, low woods, flood plains, and ditches from Nova Scotia and Minnesota, south to Virginia and Texas

Foxglove Penstemon

DESCRIPTION

Foxglove penstemon has 6- to 8-inch, shiny green leaves that form tufted rosettes. The flowerstalks can reach 5 feet and bear branched clusters of 1-inch, somewhat inflated white flowers. Each flower has two upper and three lower lips. In the wild, the spires of lovely white flowers carpet meadows in early summer, giving them an ethereal look. The stem leaves are large and prominent. The seed capsules are decorative in fall and winter. 'Husker Red' has rubyred foliage and stems and pink flowers.

GARDEN USES

Plant foxglove penstemon in beds and borders with alumroots (*Heuchera* spp.), cranesbills (*Geranium* spp.), yarrows, ornamental onions (*Allium* spp.), garden phlox (*Phlox paniculata*), and Siberian iris (*Iris sibirica*). In meadows and other informal settings, combine it with mountain mints (*Pycnanthemum* spp.), bowman's-root (*Gillenia trifoliata*), sunflower heliopsis (*Heliopsis helianthoides*), bee balm (*Monarda didyma*), ironweeds (*Vernonia* spp.), and ferns.

GROWING AND PROPAGATION

Plant in moist, rich soil in full sun or light shade. Plants spread by slow-creeping stems to form dense clumps. Divide clumps every 4 to 6 years to keep them vigorous. Plants self-sow profusely; the seedlings may become a nuisance. To prevent self-sowing, cut off capsules before the seeds ripen.

OTHER WHITE-FLOWERED *PENSTEMON* SPECIES

Smooth beardtongue (*Penstemon laevigatus*) is similar to *P. digitalis*, but the flowers are slightly smaller and may be white or pale pink. Stems reach only 2 to 3 feet in height. Plant in moist, average to rich soil in full sun or light shade. Found in meadows, open woods, and clearings from Pennsylvania, south to Florida and Alabama. Zones 5 to 9.

White-flowered penstemon (*Penstemon albidus*) has compact flowerstalks with perky ½-inch white flowers with flat faces. The 2½-inch, deep green leaves decrease in size as they climb the 6- to 14-inch stems. Plant in average to rich, sandy or loamy, well-drained soil in full sun. Found in prairies, woodland edges, and embankments. Zones 3 to 8.

Wild foxglove (*Penstemon cobaea*) bears 2-inch, inflated, white to pale purple flowers with deep violet stripes. They bloom in tight clusters on the 1- to 2½-foot flowerstalks in spring and early summer. The deep green, oval, 2½-inch leaves have toothed margins. Plant them at the edge of a woodland, in a dry prairie garden, or in a rock garden in sandy or loamy, well-drained soil in full sun or light shade. Good drainage is essential, especially in winter. Plants may be short-lived but self-sow readily. Found in open woods and prairies. Zones 5 to 9.

Penstemon hirsutus

Hairy Beardtongue

Pronunciation	PEN-steh-mon her-SOO-tus
Family	Scrophulariaceae, Figwort Family
USDA Hardiness Zones	4 to 8
Native Habitat and Range	Open woods, meadows, outcroppings, and roadside embankments from Maine and Wisconsin, south to Virginia and Kentucky

Hairy Beardtongue

DESCRIPTION

Hairy beardtongue has rosettes of 4-inch, fuzzy, oval leaves with a red outline and toothed edges. The narrow, tubular flowers are soft purple to violet. They bloom on 2- to 3-foot stems with broad leaves. They grow from fibrous roots to form evergreen rosettes. 'Pygmaeus' is a dwarf selection with dense rosettes of reddish-tinged foliage. Mature plants form broad, tight clumps with tightly packed clusters of lilac flowers on 8-inch stems.

GARDEN USES

Combine clumps of hairy beardtongue with early summer perennials such as green and gold (*Chrysogonum virginianum*), baptisias (*Baptisia* spp.), campions (*Silene* spp.), phlox, fleabanes (*Erigeron* spp.), and grasses. Place them at the front of beds and borders with spreading and mounding plants such as poppy mallow (*Callirhoe* spp.), pinks (*Dianthus* spp.), sedums, cranesbills (*Geranium* spp.), snow-in-summer (*Cerastium* spp.), and dwarf asters.

GROWING AND PROPAGATION

Plant in average to rich, moist but well-drained soil in full sun or partial shade. Established plants tolerate drought. Divide clumps in early spring or fall to control their size or for propagation. Propagate by sowing seed outdoors in fall or indoors with 4 to 6 weeks of cold, moist stratification.

MORE *PENSTEMON* SPECIES FOR AVERAGE SOILS

Slender penstemon (*Penstemon gracilis*) has ¾-inch, rose-pink to purple flowers on 2-foot stems. The narrow, 2- to 5-inch toothed leaves are covered with soft hairs. Plant in average to rich, well-drained soil in full sun or light shade. Plants grow naturally in open woods, savannas, and prairies from Wisconsin and Alberta, south to Iowa and New Mexico. Zones 3 to 8.

Small's beardtongue (*Penstemon smallii*) is a shrubby plant with 2- to 2½-foot stems that are fully covered in 1-inch, rose-purple flowers for several weeks in early spring. The glossy, broadly lancelike, 4-inch leaves have toothed edges. Plants are short-lived and must have good drainage. Plant in average, well-drained, gravely, sandy, or loamy soil in full sun or light shade. Plants grow naturally in open woods, woodland margins, and cliffs in the southern Appalachian Mountains from North Carolina to Tennessee. Zones 6 to 8.

Southern penstemon (*Penstemon australis*) has 6-inch, toothed oval leaves. Both leaves and stems are covered with soft hairs. Flowers are white to red-violet with deep rose stripes. Plants grow 2½ to 3 feet tall. Plant in average to rich, well-drained soil in full sun or light shade. Plants grow naturally in open woods, clearings, and dunes and along roadsides on the coastal plain from Virginia to Alabama. Zones 5 to 9.

Phacelia bipinnatifida

Fern-Leaved Phacelia

Pronunciation	fah-SEE-lee-uh by-pin-nah-TIFF-ih-duh
Family	Hydrophyllaceae, Waterleaf Family
USDA Hardiness Zones	4 to 9
Native Habitat and Range	Moist, deciduous woods, woodland coves, and shaded roadsides from Virginia and Illinois, south to Georgia and Arkansas

Fern-Leaved Phacelia

DESCRIPTION

Fern-leaved phacelia is a biennial that produces attractive rosettes of broad, ferny, 8- to 12-inch leaves in its first year of growth. In its second spring, it sends up deep blue-violet flowers at the top of 1- to 2-foot stems. The flowers have five fringed or ragged petals and fuzzy stamens and occur in curved, one-sided clusters.

GARDEN USES

Plant fern-leaved phacelia under trees and shrubs alone or combined with wildflowers such as celandine poppy (*Stylophorum diphyllum*), baneberries (*Actaea* spp.), wild bleeding heart (*Dicentra eximia*), Canada wild ginger (*Asarum canadense*), wild cranesbill (*Geranium maculatum*), and ferns. Plants reseed readily amid other garden plants, but they are not invasive. They will come up between clumps of larger plants such as blue cohosh (*Caulophyllum thalictroides*) and black snakeroot (*Cimicifuga racemosa*). The larger plants' expanding foliage hides the bare spots left when the phacelia dies out in summer.

GROWING AND PROPAGATION

Plant in moist, humus-rich soil in light to full shade. Since plants are biennial, they will bloom every other year from seed. Set out first-year plants for two successive years, and the plants will then reseed to maintain a perpetual bloom cycle. Plants reseed best on open, unmulched soil.

OTHER *PHACELIA* SPECIES

Fringed phacelia (*Phacelia fimbriata*) is an annual with limp stems and oval, bluntly lobed leaves. The ½-inch fringed flowers are white to pale lavender, appearing in open clusters at the tips of the 2-foot stems. Plant in moist, humus-rich soil in light to full shade. Open woods and clearings and along roadsides in the mountains of Virginia, North Carolina, and Tennessee. Zones 5 to 8.

Miami-mist (*Phacelia purshii*) is similar to *P. fimbriata* but has pointed leaf lobes. The flowers are pale blue with white centers. Plants grow 2 to 2½ feet tall. Plant in moist, humus-rich soil in light to full shade. Rich woods, clearings, and floodplains from Pennsylvania and Illinois, south to Georgia and Missouri. Zones 4 to 8.

Scorpion weed (*Phacelia dubia*) is an annual with spreading stems clothed in small, deeply lobed leaves. The white or blue flowers are unfringed and are carried in loose clusters at the ends of the stems. Plant in moist, humus-rich soil in light to full shade. Open woods, floodplains, and roadsides. Zones 5 to 8.

Phlox divaricata

Wild Blue Phlox, Woodland Phlox

Pronunciation	FLOCKS dih-var-ih-KAH-tuh
Family	Polemoniaceae, Phlox Family
USDA Hardiness Zones	3 to 9
Native Habitat and Range	Moist, deciduous woods, clearings, floodplains, and roadsides from Quebec and Minnesota, south to Georgia and Texas

DESCRIPTION

Wild blue phlox forms spreading clumps of glossy, broadly lance-shaped, evergreen leaves from creeping stems. The 1- to 1½-foot flowering stems are erect and hairy, topped by an open cluster of ¾-inch, fragrant, sky blue flowers with notched petals. Each flower has five petals and is tubular at the base, opening to form a flat-faced blossom. The stalk withers after flowering.

GARDEN USES

Choose wild blue phlox for early color in shade or wildflower gardens. Use under flowering shrubs or with wildflowers such as celandine poppy (*Stylophorum diphyllum*), wild columbine (*Aquilegia canadensis*), Virginia bluebells (*Mertensia virginica*), spring beauty (*Claytonia virginica*), Canada wild ginger (*Asarum canadense*), and Allegheny foamflower (*Tiarella cordifolia*). Spring bulbs, bleeding hearts (*Dicentra* spp.), and ferns are also good choices.

GROWING AND PROPAGATION

Plant in evenly moist, humus-rich soil in light to full shade. Cut the spent bloom stalks after flowering to keep the clumps neat. Plants form evergreen groundcovers that seldom need dividing unless they crowd other plants. To divide, lift plants after flowering. Take cuttings in late spring and early summer. They root quickly.

Wild Blue Phlox

Rabbits love to eat phlox. Powdery mildew may form a white coating on the leaves. To reduce mildew problems, keep plants evenly moist and thin clumps to improve air circulation.

CULTIVARS OF WILD BLUE PHLOX

'Clouds of Perfume' has intensely fragrant, ice blue flowers.

'Dirigo Ice' has pale blue flowers.

'Fuller's White' is a sturdy, compact plant with pure white flowers.

ANOTHER SPREADING *PHLOX* SPECIES

Creeping phlox (*Phlox stolonifera*) has creeping stems that form tight clumps 6 to 8 inches tall. Open clusters of broad-petaled, magenta to pink flowers appear in early to midspring; the bloom stalks die back after flowering. 'Blue Ridge' has lilac-blue flowers. 'Bruce's White' has white flowers. 'Pink Ridge' has mauve-pink flowers. 'Sherwood Purple' has purple-blue flowers. Plant in moist, humus-rich soil in light to full shade. Plants grow naturally in open deciduous woods, wooded slopes, and clearings from Pennsylvania and Ohio, south to Georgia. Zones 2 to 8.

Phlox subulata

Moss Phlox

Pronunciation	FLOCKS sub-you-LAH-tuh
Family	Polemoniaceae, Phlox Family
USDA Hardiness Zones	2 to 9
Native Habitat and Range	Open, sandy woods, outcroppings, ledges, and roadsides from New York and Michigan, south to North Carolina and Tennessee

Moss Phlox

DESCRIPTION

Moss phlox forms mats of bright blossoms in spring on slopes and banks and in rock gardens. The ¾-inch, pink, magenta, or white flowers cover the mounds of dense needlelike foliage and wiry stems. Each flower has five petals and is tubular at the base, opening to form a flat-faced blossom. Bloom lasts for several weeks. 'Alexander's Wild Rose' has pink flowers. 'Blue Hills' has medium blue flowers with darker centers. 'Emerald Cushion Blue' has blue flowers. 'Emerald Cushion Pink' has pink flowers and is long-blooming. 'Maiden's Blush' is pink-flushed white with a red spot in the center and is a good rebloomer. 'Snowflake' is a compact plant with white flowers.

GARDEN USES

Moss phlox is valuable on dry slopes and rock walls where few other plants grow well. It also works well at the front of a border or along walks. Plant moss phlox with prairie smoke (*Geum triflorum*), cranesbills (*Geranium* spp.), basket-of-gold (*Aurinia saxatilis*), sedums, rock cresses (*Arabis* spp.), penstemons (*Penstemon* spp.), spring bulbs, and grasses.

GROWING AND PROPAGATION

Plant in average, sandy or loamy, well-drained soil in full sun. They are tough, adaptable plants that grow for years with little care. Divide them in fall or take cuttings in spring or early summer.

OTHER SPINY-LEAVED *PHLOX* SPECIES

Prairie phlox (*Phlox pilosa*) is a low, spreading plant that grows to 1½ feet tall with 3-inch, stiff, lance-shaped leaves. Plants produce many leafy stems in open clumps that do not bear flowers. The ½- to ¾-inch, pink or white flowers are carried in flattened terminal clusters in early summer. There are several regional varieties of this species that vary in the width of the foliage and in flower size and color. Use prairie phlox in borders or in informal plantings with golden Alexanders (*Zizia* spp.), mountain mints (*Pycnanthemum* spp.), asters, goldenrods, and ornamental grasses. Plant in rich, moist soil in full sun or light shade. Found in open woods, savannas, and prairies. Zones 3 to 9.

Sand phlox (*Phlox bifida*) is a creeping plant with 2-inch, stiff, needlelike leaves and loose clusters of ⅜-inch spring flowers with deeply notched white to lavender petals. Plant in average to rich, sandy or loamy, well-drained soil in full sun. Sand phlox is perfect for the front of the border or in rock gardens in combination with butterfly weed (*Asclepias tuberosa*), alumroots (*Heuchera* spp.), yucca, verbenas, and prickly pear cactus (*Opuntia humifusa*). Found in dry open woods, barrens, and clearings. Zones 4 to 8.

Physostegia virginiana

Obedient Plant

Pronunciation	fie-so-STEE-gee-uh ver-jin-ee-AH-nuh
Family	Lamiaceae, Mint Family
USDA Hardiness Zones	3 to 9
Native Habitat and Range	Low, wet woods, bottomlands, wet prairies, and borders of marshes from Maine and Alberta, south to South Carolina and Texas

Obedient Plant

DESCRIPTION

Obedient plant, also known as Virginia false dragonhead, produces showy spikes of 1-inch, rose-pink to lilac flowers in late summer. The flowers are arranged in four vertical rows on the flower spikes; the plant gets its "obedient" label from the flowers' tendency to remain in any position to which they are moved. Pairs of coarsely toothed, lance-shaped leaves cover the 3- to 4-foot stems. Plants spread quickly from creeping stems with fibrous roots.

GARDEN USES

Obedient plant is an old-fashioned perennial popular in colonial and Victorian gardens. Choose it for streamside plantings, meadow gardens, beds, and borders. Combine the vertical spikes with late summer perennials such as ironweeds (*Vernonia* spp.), boltonia (*Boltonia asteroides*), garden phlox (*Phlox paniculata*), asters, Joe-Pye weeds (*Eupatorium* spp.), goldenrods, sunflowers, and ornamental grasses.

GROWING AND PROPAGATION

Plant in evenly moist, humus-rich soil in full sun or partial shade. Plants will grow well in fairly wet soils beside ponds and streams. The plants are heavy feeders, but you may need to stake them to keep them from flopping over if you plant them in rich soil. Divide plants every 2 to 3 years to control their spread, and replant them into amended soil, or propagate by taking stem cuttings in early summer.

CULTIVARS OF OBEDIENT PLANT

'Pink Bouquet' has bright pink flowers on 3- to 4-foot stems.

'Red Beauty' has rose-red flowers on 2½-foot stems.

'Rosea' has deep rose-pink flowers.

'Summer Snow' is pure white and compact (to 3 feet). It spreads less and blooms several weeks earlier than the other cultivars.

'Variegata' has pale pink flowers and leaves edged in creamy white.

'Vivid' has vibrant rose-pink flowers on 2- to 2½-foot stems.

Podophyllum peltatum

Mayapple

Pronunciation	poe-doe-FILL-um pell-TAY-tum
Family	Berberidaceae, Barberry Family
USDA Hardiness Zones	3 to 9
Native Habitat and Range	Rich, deciduous woods, floodplains, bottomlands, clearings, and shaded roadsides from Quebec and Minnesota, south to Florida and Texas

Mayapple

DESCRIPTION

The creamy white, 2-inch flowers of mayapple nod beneath the lush, paired leaves in early spring. The leaves remain fresh all season in cooler zones, but plants are dormant by mid- to late summer in hot, dry areas. The egg-shaped fruit is edible only when ripe in late summer. Green "apples" are poisonous, as are all other portions of the plant. Plants spread from thick, creeping, underground rhizomes.

GARDEN USES

Mayapple is a perfect choice for covering large areas of bare ground. Plant it under shrubs and trees or combine with larger wildflowers such as Solomon's seals (*Polgonatum* spp.), baneberries (*Actaea* spp.), black snakeroot (*Cimicifuga racemosa*), spikenard (*Aralia racemosa*), and ferns. It also looks lovely with early flowers such as Virginia bluebells (*Mertensia virginica*), spring beauties (*Claytonia* spp.), and Dutchman's breeches (*Dicentra cucullaria*), which go dormant before the mayapple foliage shades the ground.

GROWING AND PROPAGATION

Plant in evenly moist to damp, humus-rich soil in light to full shade. Plants spread so rapidly that they often need control, especially in small gardens. Divide plants in late summer or fall as they are going dormant.

Rust infection may cause yellow or orange spots on the leaves. To control rust, dig and destroy affected plants.

PERFECT GARDEN COMPANIONS

Skunk cabbage (*Symplocarpus foetidus*) has fleshy, deep purple-red flowers that emerge in late winter and can actually melt the snow. The leaves emerge as the flowers fade and reach up to 3 feet across. If soil becomes dry or temperatures soar, plants may go dormant by midsummer. Use the unique bold foliage of skunk cabbage where drama is needed and surround the large leaves with the smaller umbrellas of mayapple for a unique textural combination. Plants grow from deep, stout crowns with thick, fleshy roots. Mature plants are very difficult to transplant. Plant in moist to wet, humus-rich soil in full sun or full shade. Low wet woods, bottomlands, seeps, and slow streams. Zones 3 to 9.

Tassel rue (*Trautvetteria carolinense*) has petal-less flowers with fluffy, ½-inch white stamens that carry the floral show. Leafy stalks rising to 3 feet are crowned with open clusters of the foamy white flowers in late spring. The large, rounded leaves of tassel rue are deeply cut into 5 to 11 ragged lobes. This striking foliage is indispensable in the summer garden. Plant it with skunk cabbage so its persistent leaves will fill the void left when skunk cabbage goes dormant in summer. Plants grow from fibrous-rooted crowns. Plant in moist to wet, humus-rich soil in full sun or partial shade. Streamsides, seeps, wet prairies, and ditches. Zones 4 to 8.

Polemonium reptans

Creeping Jacob's Ladder

Pronunciation	po-leh-MOW-nee-um REP-tanz
Family	Polemoniaceae, Phlox Family
USDA Hardiness Zones	3 to 8
Native Habitat and Range	Moist, deciduous woods, clearings, floodplains, and shaded roadsides from New York and Minnesota, south to Virginia, Alabama, and Arkansas

Creeping Jacob's Ladder

DESCRIPTION

Creeping Jacob's ladder produces open to rounded clumps of attractive, dark green foliage on short, branching, 8- to 16-inch stems. Each leaf consists of several evenly spaced leaflets. Open clusters of ½-inch, nodding, deep sky blue flowers cover the plants in midspring. Each flower has five overlapping petals that form a cup.

GARDEN USES

Plant masses of creeping Jacob's ladder under shrubs or flowering trees such as serviceberries (*Amelanchier* spp.) or Carolina silverbell (*Halesia carolina*). Combine it with wildflowers such as bloodroot (*Sanguinaria canadensis*), wild cranesbill (*Geranium maculatum*), Allegheny foamflower (*Tiarella cordifolia*), and wild gingers (*Asarum* spp.). You can also plant it in traditional gardens with hostas, ferns, and border perennials.

GROWING AND PROPAGATION

Plant in evenly moist, humus-rich soil in full sun or partial shade. In warm regions, partial shade is a must to keep the foliage from burning. After flowering, cut the bloom stalks to the ground and the foliage will remain attractive all season. Plants seldom need division, but you can lift and divide the crowns in early spring or fall. Propagate by seeds sown outdoors in fall.

CULTIVARS OF CREEPING JACOB'S LADDER

'Blue Pearl' bears a profusion of rich, medium blue flowers.

'Lambrook Manor' has showy, outward-facing, lilac-blue flowers on 2-foot-tall plants.

A PERFECT GARDEN COMPANION

Creeping mint or **Meehan's mint** (*Meehania cordata*) is an unusual and strikingly beautiful groundcover. In time, plants form dense clumps of softly fuzzy, heart-shaped leaves with scalloped edges. The rich blue, tubular flowers are abundant and showy in spring. Combine them with shooting stars (*Dodecatheon* spp.), woodland phlox (*Phlox divaricata*), bloodroot (*Sanguinaria canadensis*), ferns, and other plants that bloom at the same time as Jacob's ladder. Plant in evenly moist, humus-rich soil in light to full shade. Plants may take 2 to 3 years before they really settle in. Once established, they spread well and are quite hardy. Take stem cuttings in early summer. Rich, deciduous woods. Zones 4 to 8.

Polygonatum biflorum

Solomon's Seal

Pronunciation	poe-lig-oh-NAY-tum bye-FLOOR-um
Family	Liliaceae, Lily Family
USDA Hardiness Zones	3 to 9
Native Habitat and Range	Rich, moist, deciduous or mixed coniferous woods, floodplains, clearings, and roadsides from Massachusetts and Manitoba, south to Florida and Mexico

Solomon's Seal

DESCRIPTION

Solomon's seal has narrowly oval, deep gray-green leaves with white undersides. The leaves are closely spaced on the graceful, arching 1- to 3-foot stems. The ½-inch, green, bell-like flowers have white tips. They hang below the foliage in pairs. Plants produce blue-black berries in late summer that attract birds. The leaves turn pure yellow in the fall.

GARDEN USES

Place Solomon's seal to tower over a low groundcover planting or against the backdrop of a tree trunk or fallen log. Combine it with wild gingers (*Asarum* spp.), bloodroot (*Sanguinaria canadensis*), mayapple (*Podophyllum peltatum*), creeping Jacob's ladder (*Polemonium reptans*), and phlox. Solomon's seal can tolerate the dry shade under mature trees: Try mass plantings in combination with ferns under shrubs and flowering trees.

GROWING AND PROPAGATION

Plant in moist to dry, humus-rich soil in partial to full shade. Plants tolerate sun in the North but must have shade from all but morning sun in the South. They spread by branching rhizomes and can quickly outgrow their site. Divide clumps in spring or fall and replant them. To propagate, remove the seeds from the pulpy fruit when it turns blue-black and sow immediately outdoors. Seeds may take 2 years to germinate, and seedlings grow slowly.

OTHER *POLYGONATUM* SPECIES

Great Solomon's seal (*Polygonatum biflorum* var. *commutatum*) has thick, 3- to 7-foot stems and well-spaced, 3- to 7-inch, oval leaves. The greenish white flowers appear in clusters of three to eight. Plant in evenly moist soil in light to full shade. Plants spread rapidly. Place these stately giants where their tall arching stems add a vertical accent to the garden. Combine them with large ferns such as ostrich fern (*Matteuccia struthiopteris*) and Goldie's fern (*Dryopteris goldiana*) and bold perennials like spikenard (*Aralia racemosa*) and baneberries (*Actaea* spp.). Rich, deciduous woods, clearings, bottomlands, and roadsides from New Hampshire and Manitoba, south to Georgia and northern Mexico. Zones 3 to 8.

Hairy Solomon's seal (*Polygonatum pubescens*) has 5-inch, narrowly oval leaves with hairy veins on the lower leaf surfaces. Yellow-green flowers are found in groups of two or three on 1- to 3-foot stems. Plant in moist or dry, humus-rich soil in partial to full shade. Rich, deciduous woods and clearings from Nova Scotia and Manitoba, south to Georgia and Indiana. Zones 3 to 8.

Pycnanthemum virginianum

Virginia Mountain Mint

Pronunciation	pick-NAN-thuh-mum ver-jin-ee-AY-num
Family	Lamiaceae, Mint Family
USDA Hardiness Zones	3 to 8
Native Habitat and Range	Open woods, meadows, and prairies from Maine and North Dakota, south to Georgia and Oklahoma

Virginia Mountain Mint

DESCRIPTION

Virginia mountain mint produces frothy, flattened mounds of small, white flowers spotted with purple from late spring to midsummer. The attractive, 2-inch, narrow leaves smell like mint and oregano. The 2- to 3-foot plants turn bright yellow in the fall, and the gray seedheads cling to the plants through the winter.

GARDEN USES

Use Virginia mountain mint in meadow and prairie plantings or in roomy beds where it can spread. Combine it with purple prairie clover (*Dalea purpurea*), gayfeathers (*Liatris* spp.), black-eyed Susans (*Rudbeckia* spp.), butterfly weed (*Asclepias tuberosa*), and phlox. Mountain mint is very attractive to butterflies. The fine-textured heads are a good foil to bold flowers such as purple coneflowers (*Echinacea* spp.), bellflowers (*Campanula* spp.), and lilies.

GROWING AND PROPAGATION

Plant in moist, humus-rich soil in full sun. Plants spread rapidly and need division every 2 years to keep them from swamping other plants. Lift clumps in early spring or fall and replant the vigorous portions. Take stem cuttings in early summer. Sow seeds indoors in spring with 4 weeks of cold, moist stratification.

OTHER *PYCNANTHEMUM* SPECIES FOR FULL SUN

Narrow-leaved mountain mint (*Pycnanthemum tenuifolium*) looks like a delicate version of *P. virginianum.* Its needlelike leaves are smaller and narrower, and plants reach only 1 to 1½ feet. Plant in rich, moist soil in full sun. Plants grow naturally in low or dry open woods, meadows, and bogs from Maine and Wisconsin, south to South Carolina and Texas. Zones 4 to 8.

Short-toothed mountain mint (*Pycnanthemum muticum*) is a stunning plant with broad clusters of dense heads with ¼-inch, white flowers accented by broad, white-frosted bracts. The pointed, oval leaves are peppermint-scented. Plants grow to 3 feet tall with an equal spread. Plant in average to rich, well-drained soil in full sun or light shade. Plants spread by creeping underground stems to form large clumps. Divide them every 3 or 4 years to control their spread and keep them vigorous. Take stem cuttings in early summer. Found in moist open woods, clearings, and meadows and along roadsides. Zones 4 to 9.

Ratibida pinnata

Gray-Headed Coneflower

Pronunciation	ruh-TIB-ih-duh pin-AH-tuh
Family	Asteraceae, Aster Family
USDA Hardiness Zones	3 to 9
Native Habitat and Range	Prairies, savannas, and roadsides and along railroad tracks from Ontario and South Dakota, south to Georgia and Oklahoma

Gray-Headed Coneflower

DESCRIPTION

The slender, 3- to 5-foot stalks of gray-headed coneflower bear open clusters of flowers with 2- to 3-inch, drooping, yellow petals and prominent brown, oval centers that turn gray in autumn. The 4- to 10-inch, pinnately lobed leaves form open basal rosettes that grow sparsely on the stems.

GARDEN USES

Use sweeps of gray-headed coneflower in meadow and prairie gardens or as accents in the middle or rear of perennial beds. Combine them with gayfeathers (*Liatris* spp.), ironweeds (*Vernonia* spp.), milkweeds (*Asclepias* spp.), anise hyssop (*Agastache foeniculum*), queen-of-the-prairie (*Filipendula rubra*), and grasses. In formal situations, use them with yarrows, phlox, verbenas (*Verbena* spp.), daisies, and bee balms (*Monarda* spp.).

GROWING AND PROPAGATION

Plant in average to rich, moist soil in full sun or light shade. Plants form tight clumps of woody, fibrous-rooted crowns that seldom need division. In a crowded planting, they may give way to more vigorous plants, but they will reseed in more favorable spots. In partial shade or in rich soils, plants may flop from the weight of the flowerheads. Sow seeds outdoors in autumn or indoors in spring with 4 weeks of cold, moist stratification. Self-sown seedlings may be plentiful.

ANOTHER RECOMMENDED *RATIBIDA* SPECIES

Mexican hat (*Ratibida columnifera*) is a diminutive version of gray-headed coneflower that grows 1 to 3 feet tall. Throughout the summer, the dense, branching stems are crowned with exaggerated, sombrero-like flowers, each composed of a 1- to 1½-inch central cone surrounded by short, 1-inch, drooping petals. Mexican hats grow best in sites where they don't have to compete with larger and more vigorous plants, and they will self-sow profusely in such conditions. Plant in average, sandy or loamy, well-drained soil in full sun. Found on dry prairies, high plains, and open woods and along roadsides from Minnesota and Montana, south to Missouri and Mexico. Zones 3 to 9.

Rudbeckia fulgida

Orange Coneflower

Pronunciation	ruhd-BECK-ee-uh FULL-jih-duh
Family	Asteraceae, Aster Family
USDA Hardiness Zones	3 to 9
Native Habitat and Range	Moist, open woods, clearings, meadows, and roadsides from Pennsylvania and Michigan, south to Florida and Texas

Orange Coneflower

DESCRIPTION

Orange coneflower is a popular perennial that bears showy, 2- to 2½-inch, orange-yellow petals surrounding deep brown centers. It blooms for a month or more in late summer. The rough, hairy leaves are oval to lance-shaped. The species includes several recognized varieties that differ in the size of the flowers and in the width of the leaves. Plants grow 1½ to 3 feet tall.

GARDEN USES

Beautiful, easy-care orange coneflowers are perfect for adding bright, long-lasting color to perennial gardens, meadows, and municipal plantings. Combine them with summer flowers such as coreopsis, Russian sage (*Perovskia atriplicifolia*), garden phlox (*Phlox paniculata*), sedums, and ornamental grasses. In meadows, plant them with bee balms (*Monarda* spp.), purple coneflowers (*Echinacea* spp.), goldenrods, and asters.

GROWING AND PROPAGATION

Plant in moist but well-drained, average to rich soil in full sun or light shade. Plants grow vigorously to form broad clumps from branched, fibrous-rooted crowns. Divide every 2 to 3 years in spring or fall to control their spread and to rejuvenate the clumps. Replant divisions in soil enriched with compost or manure. Leave flowerstalks standing for winter interest. Birds such as goldfinches and chickadees will cover the plants in fall and winter, devouring the seeds within the showy cones. Self-sown seedlings will appear.

ANOTHER RECOMMENDED *RUDBECKIA* SPECIES

Black-eyed Susan (*Rudbeckia hirta*) is the familiar yellow daisy of roadsides and meadows. They are biennials or short-lived perennials with 1- to 2½-inch flowers on 1- to 1½-foot stems. The narrow leaves are clothed in bristly hairs. Plant in average to rich, well-drained soil in full sun or light shade. Found in open woods, meadows, prairies, and waste places from Newfoundland and British Columbia, south to Florida and Mexico. The hybrid Gloriosa daisies are derived from this species. They have 5- to 6-inch yellow, orange, red, or multicolored flowers on 2- to 3-foot plants. Zones 2 to 9.

Rudbeckia laciniata

Ragged Coneflower, Green-Headed Coneflower

Pronunciation	ruhd-BECK-ee-uh luh-sin-ee-AH-tuh
Family	Asteraceae, Aster Family
USDA Hardiness Zones	3 to 9
Native Habitat and Range	Open, deciduous woods, bottomlands, floodplains, and clearings from Quebec and Montana, south to Florida and Arizona

Ragged Coneflower (yellow)

DESCRIPTION

Ragged coneflower is a tall, stout plant that has 16-inch, three- to five-lobed basal leaves and leafy stems with smaller, three-lobed leaves. The 3- to 6-foot stems are crowned by branched clusters of 2- to 3-inch, yellow flowers with drooping ray petals and a conical green center. They bloom for several weeks in mid- to late summer. 'Gold Drop' has fully double flowers on compact, 2- to 3-foot stems. 'Golden Glow' has 3½- to 4-inch double lemon yellow flowers on 3- to 5-foot stems.

GARDEN USES

Ragged coneflowers are excellent for informal meadow and prairie gardens or the rear of perennial beds. They also perform well in lightly shaded woodland gardens. Combine them with obedient plant (*Physostegia virginiana*), sneezeweeds (*Helenium* spp.), goldenrods, ironweeds (*Vernonia*. spp.), asters, grasses, and ferns.

GROWING AND PROPAGATION

Plant in moist but well-drained, average to rich soil in full sun or light shade. Plants grow vigorously in the sun but are more delicate in shade. Aphids often appear on the upper stems in early summer. They cause little injury and are easy to wash from plants with a hard stream of water. Use insecticidal soap to control damaging infestations. Divide ragged coneflowers every 3 to 5 years. Cut floppy plants to the ground after flowering. Self-sown seedlings appear.

OTHER TALL *RUDBECKIA* SPECIES

Giant coneflower (*Rudbeckia maxima*) is a striking plant with dense clumps of 2-foot, waxy, blue-green, oval leaves and thick stems to 6 feet tall. The flowers have 3-inch, drooping oval rays and 3- to 4-inch, dark brown conical centers. Plant in average to rich, moist but well-drained soil in full sun or light shade. Zones 4 to 9.

Shining coneflower (*Rudbeckia nitida*) is similar to ragged coneflower but has oval leaves without lobes. Plants in cultivation probably belong to *R. laciniata,* as the leaves are lobed, but they are listed under *R. nitida.* Zones 3 to 9.

Ruellia humilis

Wild Petunia

Pronunciation	rue-ELL-lee-uh HUE-mill-is
Family	Acanthaceae, Acanthus Family
USDA Hardiness Zones	4 to 9
Native Habitat and Range	Open woods, prairies, meadows, and clearings from Pennsylvania and Nebraska, south to Alabama and Texas

Wild Petunia

DESCRIPTION

Wild petunia grows 1 to 2 feet tall with hairy stems and 3-inch, stalkless leaves. The tubular, five-petaled, flat-faced flowers resemble those of the annual bedding plant petunias. The lavender or lilac-blue flowers are borne in pairs in the leaf axils for several weeks in summer.

GARDEN USES

The delicate wild petunia is likely to be overlooked if planted with exuberant plants. Place them at the front of the border, along a trail, at the bottom of steps, or in a rock garden. Combine them with red fire pink (*Silene virginica*), prairie smoke (*Geum triflorum*), coreopsis, ornamental onions, green and gold (*Chrysogonum virginianum*), and smaller grasses. In meadows and prairies, grow them with more delicate plants such as prairie phlox (*Phlox pilosa*), pussy toes (*Antennaria* spp.), butterfly weed (*Asclepias tuberosa*), and Virginia mountain mint (*Pycnanthemum virginicum*).

GROWING AND PROPAGATION

Plant in average to rich, moist but well-drained soil in full sun or partial shade. Wild petunia grows from fibrous-rooted crowns to form dense, attractive clumps that are never invasive. Plants may be crowded out or smothered by more exuberant neighbors, so leave them room in the garden. Take tip cuttings in early summer. Rooted offsets are easily removed from the clump in autumn. Sow seeds outdoors in autumn or indoors in spring.

OTHER *RUELLIA* SPECIES

Carolina wild petunia (*Ruellia caroliniensis*) is larger than *R. humilis*, sometimes reaching 2 feet, with 5-inch-stalked leaves covered in dense, soft hairs. The 2-inch flowers are lilac to lavender blue. Plant in moist, rich soil in full sun or partial shade. Found in open woods and clearings from New Jersey and Indiana, south to Florida and Texas. Zones 4 to 9.

Violet wild petunia (*Ruellia nudiflora*) is an upright to sprawling plant with leafy stems to 2 feet tall. The lavender to purple flowers are carried in open clusters in spring and sporadically throughout the year. Plant in sandy loam or clay soils in full sun or light shade. Found in open woods, on prairies, and along roadsides in Texas and adjacent areas of Mexico. Zones 8 to 11.

Sanguinaria canadensis

Bloodroot

Pronunciation	san-gwen-AIR-ee-uh can-uh-DEN-sis
Family	Papaveraceae, Poppy Family
USDA Hardiness Zones	3 to 9
Native Habitat and Range	Moist, deciduous woods, woodland coves, floodplains, and rocky slopes from Nova Scotia and Manitoba, south to Florida and Oklahoma

DESCRIPTION

Bloodroot produces a single leaf with seven distinct lobes. The leaf emerges wrapped around the single flower bud, which opens to a 2-inch, snow white flower with 8 to 11 narrow petals surrounding a cluster of yellow-orange stamens. Flowers open in early to midspring but last only a few days. The attractive foliage persists through the season. 'Multiplex' has long-lasting double flowers and huge, thick leaves. Plants grow from a thick, creeping rhizome.

'Multiplex', a cultivar of bloodroot

Bloodroot

GARDEN USES

Plant bloodroot as a groundcover under shrubs or in shade and woodland gardens. Combine it with spring beauties (*Claytonia* spp.), merrybells (*Uvularia* spp.), creeping Jacob's ladder (*Polemonium reptans*), Virginia bluebells (*Mertensia virginica*), and ferns. Spring bulbs, hostas, and primroses are also good companions.

GROWING AND PROPAGATION

Plant bloodroot in moist, humus-rich soil in light to full shade. It blooms best if it gets spring sun but will not survive without shade in summer. Plants go dormant during long dry spells. To propagate, divide the dense clumps in the fall, or sow fresh seeds outdoors in summer. Self-sown seedlings will appear.

A PERFECT GARDEN COMPANION

Wreath goldenrod or **blue-stemmed goldenrod** (*Solidago caesia*) is a spiky, wand-flowered goldenrod with blue-green, linear leaves alternating up wiry 1- to 3-foot stems. The leaves grow amid the flowers but droop below the stem while the flowers face upward. In early summer, the fine-textured foliage combines well with the bold leaves of bloodroot. The bright yellow flowers begin to bloom in early fall and last for several weeks. Plant in rich, moist soil in light to full shade. Plants tolerate dry soil. Open woods, woodland borders, and clearings. Zones 4 to 9.

Sanguisorba canadensis

Canadian Burnet

Pronunciation	san-gweh-SORE-buh can-uh-DEN-sis
Family	Rosaceae, Rose Family
USDA Hardiness Zones	3 to 8
Native Habitat and Range	Wet meadows and prairies, marsh and bog margins, areas near springs, and low woods from Newfoundland and Manitoba, south to New Jersey and Indiana, and in the mountains to North Carolina

Canadian Burnet

DESCRIPTION

Canadian burnet produces thick stems 4 to 5 feet tall and forms dense clumps at maturity. Each leaf is composed of several oblong, toothed leaflets. In late summer, dense spikes of fuzzy, white flowers appear. Plants grow from thick, fleshy crowns.

GARDEN USES

Plant Canadian burnet at the back of the border with phlox, monkshoods (*Aconitum* spp.), asters, goldenrods, boltonia (*Boltonia asteroides*), Joe-Pye weeds (*Eupatorium* spp.), sneezeweeds (*Helenium* spp.), and grasses. It grows well along streams or at the edges of ponds with ironweeds (*Vernonia* spp.), monkey flowers (*Mimulus* spp.), hibiscus, and ferns. In low meadows and bog gardens combine Canadian burnet with tassel rue (*Trautvetteria carolinense*), false hellebore (*Veratrum viride*), turks-cap lily (*Lilium superbum*), boneset (*Eupatorium perfoliatum*), and cow parsnip (*Heracleum lanatum*).

GROWING AND PROPAGATION

Plant Canadian burnet in evenly moist, humus-rich soil in full sun to partial shade. Plants will not tolerate high heat. Mulch plants to help keep the soil cool and moist. Divide overgrown clumps in spring. Sow seeds outdoors in fall.

A PERFECT GARDEN COMPANION

Umbrella leaf (*Diphylleia cymosa*) has huge, rounded leaves with deeply cut, pointed lobes. The stem leaves are similar but are pinched in the middle and give the illusion of a pair of wings. The small white flowers bloom in terminal clusters in spring. In late summer, the berries turn deep purple on brilliant red stalks. Mature plants have many stems. The tall spikes of Canadian burnet are an attractive foil to umbrella leaf's bold foliage, while the berries of umbrella leaf appear with the flowers of Canadian burnet. The plants often grow together in shaded woods. In warmer zones, plants go dormant in late summer. Plant in evenly moist, humus-rich soil in partial to full shade. Zones 4 to 7.

Scutellaria serrata

Showy Skullcap

Pronunciation	skew-teh-LARE-ee-uh sair-RAY-tah
Family	Lamiaceae, Mint Family
USDA Hardiness Zones	4 to 8
Native Habitat and Range	Rich, open, deciduous woods, clearings, woodland borders, and roadsides from Pennsylvania and Ohio, south to North Carolina and Tennessee, mostly in the mountains

Showy Skullcap

DESCRIPTION

Showy scullcap forms a bushy, 1- to 2-foot clump and has oval, toothed leaves. Dense spikes of showy, 1-inch, medium blue flowers bloom for months in summer. Each flower has an arching upper lip and a broader, speckled lower lip. 'Blue on Blue' has bicolored flowers, with the upper lip pale blue and the lower lip deep blue. 'Mountain Mist' is a vigorous grower with powder blue flowers. 'Whitecaps' has glistening white flowers.

GARDEN USES

Plant drifts of showy scullcaps along a woodland path or in a lightly shaded border. Combine them with bowman's-root (*Gillenia trifoliata*), Indian pink (*Spigelia marilandica*), green and gold (*Chrysogonum virginianum*), rue anemone (*Anemonella thalictroides*), Solomon's plume (*Smilacina* spp.), bloodroot (*Sanguinaria canadensis*), phlox, irises, and ferns. In rock gardens, combine them with prairie phlox (*Phlox pilosa*), pussy toes (*Antennaria* spp.), hardy geraniums, alumroots (*Heuchera* spp.), and wild bleeding heart (*Dicentra eximia*).

GROWING AND PROPAGATION

Plant in moist, humus-rich soil in light to partial shade. Plants form neat, multistemmed clumps that seldom need division. Take cuttings in early summer or sow seeds outdoors in fall.

OTHER *SCUTELLARIA* SPECIES FOR SHADE

Bushy skullcap (*Scutellaria resinosa*) is a shrubby plant that boasts a profusion of ¾-inch, deep blue flowers for months in summer. The thick, gray-green, oval leaves are decorative when the plant is out of bloom. Many tightly packed stems rise to 10 inches from a woody, tap-rooted crown. Plants demand average to lean, well-drained soil in full sun. Set out young transplants and do not disturb established clumps. Sow seeds in January for transplants by summer. Dry, rocky prairies and high plains. Zones 4 to 8.

Downy skullcap (*Scutellaria incana*) is a tall plant to 4 feet with branched terminal clusters of 1-inch, deep blue flowers. The pointed oval leaves stand out at right angles to the stems. Plant in rich, moist but well-drained soil in full sun or partial shade. Open woods and clearings and along roadsides from New York and Wisconsin, south to Virginia and Kansas, and in the mountains to Georgia. Zones 4 to 8.

Hyssop skullcap (*Scutellaria integrifolia*) is a delicate plant with 1- to 2-inch, lance-shaped leaves on 1½- to 2-foot stems. Open, terminal clusters of 1-inch, pale to medium blue flowers bloom in late spring and early summer. Plant in average to rich, moist but well-drained soil in full sun or partial shade. Meadows, clearings, open woods, and dunes. Zones 5 to 9.

Sedum ternatum

Whorled Stonecrop, Wild Stonecrop

Pronunciation	SEE-dum ter-NAH-tum
Family	Crassulaceae, Sedum Family
USDA Hardiness Zones	4 to 8
Native Habitat and Range	Moist, deciduous woods, floodplains, outcroppings, ledges, and streamsides from New England and Michigan, south to Georgia and Arkansas

DESCRIPTION

Whorled stonecrop is an open, spreading plant from 2 to 6 inches tall with thick, fleshy, waxy leaves. The flower clusters are large, and the ½-inch, white flowers have narrow, pointed petals and fuzzy stamens. 'White Waters' has larger flowers and is quite showy. Plants bloom in spring or summer.

GARDEN USES

Use whorled stonecrop as a groundcover under trees. In rock gardens and open woodlands, combine it with spring bulbs, toothwort (*Dentaria* spp.), crested iris (*Iris cristata*), Meehan's mints (*Meehania* spp.), merrybells (*Uvularia* spp.), and ferns. Use it to hold a moist, shaded bank where few other plants can become established. It will grow well between pavers on a shaded terrace or along an informal pathway. Whorled stonecrop is charming when planted in an old stump and allowed to cascade down the sides. The plants also do well in traditional containers.

GROWING AND PROPAGATION

Plant in rich, evenly moist but well-drained soil in full sun to partial shade. Plants require little care once established, and spread to form broad, dense mats. Divide in spring or fall to control their spread. Cuttings root readily throughout the summer.

Whorled Stonecrop

ANOTHER *SEDUM* SPECIES

Allegheny stonecrop (*Sedum telephioides*) has 4- to 6-inch creeping stems with lance-shaped, evergreen leaves. The plant bears domed clusters of white flowers in early summer. Plant in average to rich, well-drained soil in full sun or partial shade. Plants grow naturally on outcroppings and cliffs and in rocky woods in the mountains from Pennsylvania, south to Georgia. Zones 5 to 8.

A PERFECT GARDEN COMPANION

Wild oats or **sessile-leaved bellwort** (*Uvularia sessilifolia*) has slender stems that rise 8 to 12 inches above the soil. The stalkless, 3-inch, narrowly oval leaves are light green above and pale beneath. The 1-inch, pale straw yellow flowers bloom in late spring. Plants spread by creeping rhizomes to form broad, open clumps. Choose wild oats for foliage effect as a mass planting. The delicate individual flowers are best appreciated at close range. The plants form an airy groundcover combined with whorled stonecrop and other woodland wildflowers such as bloodroot (*Sanguinaria canadensis*), Jacob's ladder (*Polemonium reptans*), and Virginia bluebells (*Mertensia virginica*). Plant wild oats in rich, evenly moist, acid soil in partial to full shade. Plants seldom need division. Plants grow naturally in low deciduous or mixed coniferous woods in moist acid soils. Zones 3 to 8.

Senecio aureus

Golden Ragwort, Golden Groundsel

Pronunciation	seh-NEE-see-oh AW-ree-us
Family	Asteraceae, Aster Family
USDA Hardiness Zones	3 to 9
Native Habitat and Range	Low, deciduous woods, bottomlands, floodplains, meadows, and areas near springs from Labrador and Minnesota, south to Georgia and Arkansas

Golden Ragwort

DESCRIPTION

Golden ragwort has lush rosettes of deep green, heart-shaped leaves. In late spring, 1½- to 2-foot bloom stalks arise, bearing ragged-looking leaves and open, domed clusters of ¾-inch, golden yellow flowers with green or yellow centers. Long after the flowers fade, the attractive foliage persists as a neat groundcover that turns yellow in the fall.

GARDEN USES

Golden ragwort adds a touch of sunshine to the spring garden. In meadows and wild gardens, plant it with wild blue phlox (*Phlox divaricata*), wild columbine (*Aquilegia canadensis*), and Allegheny foamflower (*Tiarella cordifolia*). In boggy soil, combine it with purple avens (*Geum rivale*), skunk cabbage (*Symplocarpus foetidus*), blue flag iris (*Iris versicolor*), Jack-in-the-pulpit (*Arisaema triphyllum*), violets, and ferns. In formal beds and borders, combine it with bleeding hearts (*Dicentra* spp.), daffodils, lungworts (*Pulmonaria* spp.), Virginia bluebells (*Mertensia virginica*), and primroses.

GROWING AND PROPAGATION

Plant in rich, evenly moist to wet soil in full sun or partial shade. Plants spread readily by creeping roots and self-sown seedlings. Lift and divide clumps after flowering or in the fall. Sow seeds indoors with 4 weeks of cold, moist stratification.

MORE *SENECIO* SPECIES

Prairie ragwort (*Senecio plattensis*) has arrow-shaped basal leaves and ragged stem leaves on 1- to 1½-foot stems topped with tight clusters of ½-inch, bright yellow flowers. Plant in rich, well-drained soil in full sun or light shade. It is biennial but will self-sow. Open woods, savannas, and prairies and along roadsides. Zones 3 to 9.

Small's ragwort (*Senecio smallii*) is more robust than prairie ragwort. The flowers bloom in domed clusters on 2- to 2½-foot stems. Plant in average to rich soil in full sun or light shade. Meadows, pastures, clearings, and ditches from Pennsylvania, south to Kentucky and Florida, mostly in the mountains. Zones 4 to 9.

Shortia galacifolia

Oconee Bells

Pronunciation	SHORT-ee-uh gay-lass-ih-FOE-lee-uh
Family	Diapensiaceae, Diapensia Family
USDA Hardiness Zones	5 to 8
Native Habitat and Range	Rich, mixed woods, woodland coves, and streamsides in the foothills of the Blue Ridge mountains in North Carolina, South Carolina, and Georgia

Oconee Bells

DESCRIPTION

Oconee bells are named for their nodding, bell-like flowers with five ragged, overlapping petals. The round, glossy evergreen leaves turn purple or red in the fall. Plants form dense mats of creeping woody stems. Oconee bells are the lost flower of the Carolinas. They were discovered by French botanist Andre Michaux in 1788 in the foothills of the North Carolina mountains but could not be relocated due to Michaux's sketchy notes. The plant remained a mystery for 90 years until it was accidentally rediscovered and sent to American botanist Asa Gray for identification in 1877.

GARDEN USES

Oconee bells form an enchanting carpet under rhododendrons, azaleas (*Rhododendron* spp.), hobblebush (*Viburnum alnifolium*), and other shrubs. Try planting them along a fallen log or at the base of a boulder. Combine them with galax (*Galax urceolata*), partridgeberry (*Mitchella repens*), evergreen wild gingers (*Asarum* spp.), Allegheny foamflower (*Tiarella cordifolia*), Allegheny pachysandra (*Pachysandra procumbens*), and ferns.

GROWING AND PROPAGATION

Plant in evenly moist, humus-rich, acid soil in light to full shade. Plants spread to form broad, dense clumps. Divide them after flowering in late spring. Plants seldom self-sow.

PERFECT GARDEN COMPANIONS

Arrow-leaf wild ginger (*Asarum arifolium*), also called little brown jugs, has a unique look with 4- to 6-inch, arrow-shaped, gray-green leaves mottled with silver. The inconspicuous, light brown flowers are shaped like moonshine jugs. Use their variegated foliage in shaded gardens with Oconee bells to enhance the groundcover display. Plant in moist to dry, humus-rich soil in partial to full shade. Zones 4 to 9.

Galax, wandflower, or **beetleweed** (*Galax urceolata,* formerly listed as *G. aphylla*) has round, evergreen leaves with scalloped edges that grow in loose clusters from creeping woody rhizomes. Plants form large colonies that sport erect, 1- to 2½-foot slender wands of creamy white flowers in early summer. The leaves turn burgundy in fall. Galax is the perfect groundcover for shaded sites with intensely acidic soil. Plant in moist, humus-rich, acid soil in light to full shade. Plants bloom better with more light. Set out young rooted cuttings and water them well the first season. The plants are slow to take hold, but established plants spread well. Zones 4 to 8.

Virginia wild ginger (*Asarum virginicum*) has broad, heart-shaped leaves mottled with white. The spotted, purple, juglike flowers are 1 inch long with wide flaring lobes. Give plants moist, humus-rich soil in partial to deep shade. Zones 4 to 8.

Silene virginica

Fire Pink

Pronunciation	sigh-LEE-nee ver-JIN-ih-kuh
Family	Caryophyllaceae, Pink Family
USDA Hardiness Zones	4 to 8
Native Habitat and Range	Open woods, clearings, meadows, and roadsides from New Jersey and Ontario, south to Georgia and Oklahoma

Fire Pink

DESCRIPTION

Fire pink produces a haze of brilliant red flowers on 2- to 3-foot, wiry stems in late spring and summer. The 1-inch flowers have narrow, notched petals. Plants have a basal rosette of deep green, narrow leaves.

GARDEN USES

Use fire pink at the front of beds, borders, and rock gardens. Combine it with green and gold (*Chrysogonum virginianum*), phlox, crested iris (*Iris cristata*), bowman's-root (*Gillenia trifoliata*), coreopsis, blue-eyed grass (*Sisyrinchium* spp.), wild petunias (*Ruellia* spp.), and grasses.

GROWING AND PROPAGATION

Plant in average to rich, well-drained soil in full sun or light shade. Set out young plants and do not disturb them once they are established. Plants may be short-lived in gardens, but they self-sow readily. Sow seeds outdoors in fall or indoors uncovered with 4 weeks of cold, moist stratification. Seedlings grow quickly.

ANOTHER *SILENE* SPECIES

Wild pink (*Silene caroliniana*) is similar to fire pink, but the stems tend to grow horizontally and the flowers are clear, medium pink. The tufted leaves resemble grass. Plant in rich, well-drained soil in full sun or partial shade. Zones 5 to 8.

ANOTHER RED-FLOWERED *SILENE* SPECIES

Royal catchfly (*Silene regia*) is 2 to 5 feet tall and its leaves are attached directly to the stems. Clusters of 1-inch, fiery red flowers bloom at the tops of the stems. Plant in average to rich, well-drained soil in full sun or light shade. Plants form a deep taproot. Open woods, glades, clearings, and prairies from Ohio and Missouri, south to Georgia and Alabama. Zones 4 to 8.

PERFECT GARDEN COMPANIONS

Creeping goldenrod (*Solidago sphacelata*) is a ground-hugging goldenrod with deep green, paddle-shaped leaves and 1- to 2-foot inflorescences with stiff, horizontal branches. The low mounds of foliage complement fire pink's early flowers. Plant in average to rich, moist soil in full sun or light shade. Zones 4 to 9.

Mountain phlox (*Phlox ovata*) is an upright, spreading phlox that has glossy, oval leaves and open clusters of white, pink, or magenta flowers in late spring. Plants grow 1 to 1½ feet tall. They form large drifts in open woods and on roadsides with fire pink. Plant in moist, rich soil in full sun or partial shade. Zones 4 to 8.

Silphium laciniatum

Compass Plant

Pronunciation	SILL-fee-um lah-sin-ee-AY-tum
Family	Asteraceae, Aster Family
USDA Hardiness Zones	4 to 8
Native Habitat and Range	Dry to moist, black-soil prairies and savannas from Ohio and Minnesota, south to Alabama and Oklahoma

Compass Plant

DESCRIPTION

The towering stalks of compass plant bear huge 5-inch flowers that resemble sunflowers. Open clusters of flowers are held on stiff branches up to 8 feet above the ground. Mature clumps may have a dozen bloom stalks. The deeply lobed, 2-foot leaves form dense, decorative tufts. The leaves that rise up the stalk are smaller and less deeply cut. Plants grow from branched taproots that may grow 6 feet or more into the soil. Goldfinches, chickadees, and sparrows enjoy the large seeds.

GARDEN USES

Plant compass plant in the middle, not the back, of perennial borders, where it's easy to see both the attractive basal foliage and flowers. Combine single plants or small groups of plants with phlox, yarrows, Culver's root (*Veronicastrum virginicum*), spiderworts (*Tradescantia* spp.), poppy mallows (*Callirhoe* spp.), rattlesnake master (*Eryngium yuccifolium*), asters, and grasses. In prairie and meadow plantings they create a bold accent as they rise above the smaller flowers and grasses around them.

GROWING AND PROPAGATION

Plant in moist, well-drained, humus-rich soil in full sun. Plants need plenty of room to spread, so space them 3 to 4 feet apart. Established clumps are impossible to divide. Sow seeds outdoors in the fall or indoors in the winter with 4 weeks of cold, moist stratification. Plants will self-sow.

MORE *SILPHIUM* SPECIES WITH LARGE BASAL LEAVES

Prairie dock (*Silphium terebinthinaceum*) has a rosette of 3-foot, heart-shaped leaves. The 6- to 8-foot, leafless bloom stalks bear branched clusters of 3-inch flowers. The leaves are attractive both in the garden and in dried arrangements. Plant in rich, moist soil in full sun or light shade. Found in moist to wet prairies and savannas from Ontario and Wisconsin, south to Georgia and Louisiana. Zones 3 to 9.

Silphium compositum is similar to *S. terebinthinaceum*, but the leaves grow to only 2 feet and the flowers are 2 inches wide. There are several varieties of this species, one of which has deeply lobed leaves that resemble an oak. Plant in average to rich, well-drained soil in full sun or light shade. Found in open woods and meadows and along roadsides from Virginia, south to South Carolina and Alabama. Zones 4 to 9.

Silphium perfoliatum

Cup Plant

Pronunciation	SILL-fee-um per-foe-lee-AY-tum
Family	Asteraceae, Aster Family
USDA Hardiness Zones	3 to 8
Native Habitat and Range	Low, open woods, wet meadows, wet prairies, and ditches from Ontario and South Dakota, south to Georgia and Louisiana

Cup Plant

DESCRIPTION

The leaves of cup plant encircle the tall stems, forming a cup that collects dew. The 3-inch, yellow flowers resemble sunflowers and bloom in branched clusters just above the leaves in mid- to late summer. Plants also have a rosette of leaves at the base. Plants grow 3 to 8 feet tall. Plants form branched taproots that may grow 6 feet or more into the soil. Goldfinches visit the clumps in my garden every day, waiting for the tasty seeds to ripen.

GARDEN USES

Cup plant has a commanding presence in a formal bed, at the edge of a woodland, or in a prairie. Combine it with purple coneflowers (*Echinacea* spp.), phlox, bee balms (*Monarda* spp.), Joe-Pye weeds (*Eupatorium* spp.), asters, goldenrods, sunflowers, ironweeds (*Vernonia* spp.), and grasses.

GROWING AND PROPAGATION

Plant in rich, evenly moist soil in full sun or light shade. Young plants grow quickly to form tall, broad clumps; space plants 3 to 4 feet apart. Sow seeds outdoors in fall or indoors in spring with 4 weeks of cold, moist stratification. Plants may self-sow heavily.

Aphids may attack new growth in early summer, causing deformed leaves and stems. Spray plants with a strong stream of water to knock the pests off or use insecticidal soap to control severe infestations.

MORE *SILPHIUM* SPECIES WITH LEAFY STEMS

Rosinweed (*Silphium integrifolium*) has 2- to 4-foot stems and oval or broadly lance-shaped, toothed or untoothed leaves that are attached directly to the stems. Domed clusters of showy, 2-inch flowers bloom at the tops of the stems. Plant in average to rich, moist or dry soil in full sun or light shade. Plants grow naturally in prairies and savannas from Ohio and Wisconsin, south to Mississippi and Oklahoma. Zones 3 to 9.

Starry rosinweed (*Silphium asteriscus* var. *laevicaule,* also listed as *Silphium dentatum*) has 2- to 5-foot stems clothed in sharply toothed, narrow oval leaves. The 2-inch flowers have yellow petals that get wider at the tips, giving the flower a starry appearance. Plant in rich, well-drained soil in full sun or partial shade. Plants grow naturally in open mixed woods, savannas, meadows, and clearings from North Carolina, south to Florida and Mississippi. Zones 4 to 9.

Sisyrinchium angustifolium

Blue-Eyed Grass

Pronunciation	sis-ih-RING-key-um an-gus-tih-FOE-lee-um
Family	Iridaceae, Iris Family
USDA Hardiness Zones	3 to 9
Native Habitat and Range	Open woods, clearings, meadows, and roadsides from Newfoundland and Ontario, south to Virginia and Indiana

Blue-Eyed Grass

DESCRIPTION

Blue-eyed grass is not a true grass but a slender-leaved member of the iris family. Tufts of flat, grasslike foliage up to 20 inches tall rise from short creeping rhizomes that branch profusely. The ½-inch, steel-blue flowers have three petals and three petallike sepals and bloom in open clusters at the tops of flowerstalks. Flower color is variable, and occasionally a white-flowered plant appears. Plants bloom most profusely in late spring and early summer but may bloom off and on throughout the growing season.

GARDEN USES

Blue-eyed grass is small and delicate, so plant it in groups of three to five plants in meadows or as an accent in perennial gardens. Combine it with green and gold (*Chrysogonum virginianum*), fire pink (*Silene virginica*), phlox, prairie smoke (*Geum triflorum*), and violets. In formal gardens, use it with showy foliage plants such as silvery artemisias (*Artemisia* spp.), lamb's-ears (*Stachys byzantina*), cranesbills (*Geranium* spp.), and fine-textured ornamental grasses.

GROWING AND PROPAGATION

Plant in moist, average to rich soil in full sun or partial shade. Plants are easy to grow and spread slowly to form thick clumps. Divide clumps after flowering to reduce their size or for propagation. Self-sown seedlings will appear, often at some distance from the parent plants.

MORE *SISYRINCHIUM* SPECIES

Common blue-eyed grass (*Sisyrinchium montanum*) forms dense clumps that reach 2 feet tall and has violet flowers. Plant in average to rich, well-drained soil in full sun or light shade. Open woods, meadows, and clearings from Quebec and British Columbia, south to North Carolina and Nebraska. Zones 4 to 8.

Prairie blue-eyed grass (*Sisyrinchium campestre*) has dense, showy flower clusters. Plants grow to 16 inches tall. Plant in average to rich, well-drained soil in full sun or light shade. Open woods, savannas, prairies, and meadows from Illinois and Minnesota, south to Arkansas and Nebraska. Zones 3 to 8.

Smilacina racemosa

Solomon's Plume, False Solomon's Seal

Pronunciation	smy-lah-SEE-nuh ray-sih-MOW-suh
Family	Liliaceae, Lily Family
USDA Hardiness Zones	3 to 8
Native Habitat and Range	Rich, deciduous or mixed coniferous woods, rocky wooded slopes, clearings, and roadsides from Nova Scotia and British Columbia, south to Georgia, Arizona, and California

DESCRIPTION

Solomon's plume has 5- to 9-inch, deep green, satiny leaves and bears branching plumes of small, fuzzy white flowers in late spring. The pointed oval leaves are larger at the base of the stems and smaller at the tips. Plants grow 2 to 5 feet tall depending on age and soil fertility. The red-and-white speckled fruit is edible.

GARDEN USES

Solomon's plume is an elegant choice for informal plantings and woodland gardens. Or, plant it as a groundcover under trees and shrubs. Combine it with baneberries (*Actaea* spp.), Allegheny foamflower (*Tiarella cordifolia*), bloodroot (*Sanguinaria canadensis*), Virginia bluebells (*Mertensia virginica*), and ferns. On a shaded terrace or along a path, use it with foliage plants such as hostas, ferns, lungworts (*Pulmonaria* spp.), and epimediums (*Epimedium* spp.).

GROWING AND PROPAGATION

Plant in moist, humus-rich, nonalkaline soil in light to full shade. Plants will bloom more if they get some sun. Solomon's plume is easy to grow and forms thick clumps from branching rhizomes. Divide plants in early spring or fall for propagation or to control their spread. Sow seeds outdoors in fall.

Slugs may chew ragged holes in leaves. To stop slugs, surround your plants with a barrier strip of diatomaceous earth, or set out shallow pans of beer to trap the pests.

Solomon's Plume

ANOTHER *SMILACINA* SPECIES

Starry Solomon's plume (*Smilacina stellata*) has oblong, gray-green leaves and 2-inch, unbranched clusters of starry white flowers. The fruit has green and black stripes. Plants spread rapidly to form broad, dense clumps. Plant in average to rich, moist soil in full sun or partial shade. Plants grow naturally in open woods and on savannas, prairies, and roadsides from Newfoundland and British Columbia, south to Virginia, Missouri, and California. Zones 2 to 7.

A PERFECT GARDEN COMPANION

Starry chickweed (*Stellaria pubera*) shares its woodland home with Solomon's plume. Plant them together in shaded gardens with spring wildflowers and ferns. The low mounds of narrowly oval foliage are smothered with 1-inch starry white flowers in spring. Zones 4 to 8.

Solidago rigida

Stiff Goldenrod

Pronunciation	sole-ih-DAY-go RIJ-ih-duh
Family	Asteraceae, Aster Family
USDA Hardiness Zones	3 to 9
Native Habitat and Range	Dry to moist, gravel or black-soil prairies, meadows, clearings, and roadsides from Connecticut and Saskatchewan, south to Georgia and New Mexico

Stiff Goldenrod

DESCRIPTION

Stiff goldenrod is an oddity among goldenrods because it has large, flattened flower clusters rather than flower plumes. It forms erect, multistemmed clumps 2 to 5 feet tall. The oval leaves are covered with soft hairs. Plants bloom in late summer, and the foliage is attractive all season, turning dusty rose in the fall.

GARDEN USES

Use stiff goldenrod in beds and borders as well as in meadows and prairies. Plants are attractive in bloom as well as before they bloom. The stiff, soft green shoots add a vertical accent to the early summer garden. In formal settings, combine them with New England aster (*Aster novae-angliae*), border phlox (*Phlox paniculata*), balloon flower (*Platycodon grandiflorus*), turtleheads (*Chelone* spp.), Joe-Pye weeds (*Eupatorium* spp.), and ornamental grasses. In informal settings, combine them with asters, sunflowers, yarrows, gayfeathers (*Liatris* spp.), purple coneflowers (*Echinacea* spp.), phlox, and grasses.

GROWING AND PROPAGATION

Plant in average to lean, well-drained soil in full sun. Plants will flop if planted in shade and overly rich soil. Established plants tolerate heat and cold. Propagate by division in spring or after flowering. Sow seeds indoors in winter for transplants by summer. Plants will self-sow.

OTHER *SOLIDAGO* SPECIES WITH FLAT FLOWER CLUSTERS

Ohio goldenrod (*Solidago ohioensis*) is noteworthy for its flat-topped clusters of large, lemon yellow flowers that bloom in late summer and fall on leafy 2- to 3-foot stems. The smooth, lance-shaped basal leaves are bright green; the stem leaves overlap one another as they climb the stems. Plant in average to rich, moist soil in full sun or light shade. Plants grow naturally in bogs, wet meadows, and moist or dry prairies from Ontario and Minnesota, south to New York and Missouri. Zones 3 to 8.

Showy goldenrod (*Solidago speciosa*) forms tight clumps of leafy, red-tinged stems crowned by dense, elongated flower clusters. Plants bloom in late summer. Plant in average, sandy or loamy, well-drained soil in full sun. Plants grow naturally in prairies, meadows, savannas, and open woods from New England to Minnesota and Wyoming, south to Georgia and Texas. Zones 3 to 8.

Solidago rugosa

Rough-Stemmed Goldenrod

Pronunciation	sole-ih-DAY-go rew-GO-suh
Family	Asteraceae, Aster Family
USDA Hardiness Zones	4 to 9
Native Habitat and Range	Open woods, meadows, and old fields from Newfoundland and Michigan, south to Florida and Texas

Rough-Stemmed Goldenrod

DESCRIPTION

Rough-stemmed goldenrod has very broad, open flower clusters with arching branches in the fall. The leafy stems grow from 1 to 7 feet tall. The buds produce a bright yellow-green haze before the flowers open.

GARDEN USES

Choose rough-stemmed goldenrod for informal meadows and prairies where its tendency to spread will not be a problem. 'Fireworks' is an excellent clump-forming cultivar to plant in beds and borders with asters, sunflowers, phlox, anemones, and grasses.

GROWING AND PROPAGATION

Plant in average to rich, moist soil in full sun or light shade. Divide as needed to control spread. Rust fungi may infect the foliage, forming orange bumps.

CULTIVARS OF HYBRID GOLDENROD

'Baby Gold' is 2½ feet tall with large flower clusters.

'Cloth of Gold' has pale lemon yellow flowers on compact 1½- to 2-foot stems.

'Goldenmosa' has cascading flower clusters and lemon yellow flowers on 2½-foot stems.

OTHER *SOLIDAGO* SPECIES

Early goldenrod (*Solidago juncea*) is the first goldenrod to bloom, with flowers appearing in late July or early August. The flower clusters resemble yellow fireworks. Plants have a dense basal rosette of deep green leaves as well as leafy flowerstalks. Plant in sandy or loamy, acid soil in full sun. Open woods, meadows, savannas, lakeshores, and roadsides. Zones 3 to 8.

Gray goldenrod (*Solidago nemoralis*) produces tight clumps of gray-green basal leaves and 6- to 24-inch arching stems with one-sided plumed clusters of lemon yellow flowers that bloom from late summer to fall. Plant in lean, sandy or loamy soil in full sun or light shade. Dry meadows and prairies, savannas, and dunes. Zones 2 to 9.

Seaside goldenrod (*Solidago sempervirens*) has large, bluntly rectangular basal leaves and lance-shaped stem leaves that are smaller near the ends of the 1½- to 8-foot stems. The showy, one-sided plumes have larger flowers than most species. Plant in lean, well-drained soil in full sun. Coastal dunes and open woods. Zones 4 to 11.

Spigelia marilandica

Indian Pink

Pronunciation	spy-JEEL-ee-uh mar-ih-LAN-dih-kuh
Family	Loganiaceae, Logania Family
USDA Hardiness Zones	4 to 9
Native Habitat and Range	Moist, open woods, woodland borders, and roadsides from North Carolina and Oklahoma, south to Florida and Texas

DESCRIPTION

Indian pink is a little-known and choice plant that deserves more attention. The unusual flowers are arranged in an upright row on the one-sided inflorescence like tubes of brilliant red lipstick. The bright red, 2-inch buds open to tubular, five-petaled, lime green flowers; bloom lasts for several weeks in midsummer. Stems grow from 1 to 2½ feet tall and have pairs of glossy, deep green triangular leaves. Plants grow from fibrous-rooted crowns.

GARDEN USES

Indian pink is striking as an accent plant or in mass plantings. Use it in combination with shrubs, in borders, or along a meadow walk. Plant it with alumroots (*Heuchera* spp.), phlox, bowman's-root (*Gillenia trifoliata*), green and gold (*Chrysogonum virginianum*), gauras (*Gaura* spp.), milkweeds (*Asclepias* spp.), and coreopsis. Use it in rock gardens with yuccas (*Yucca* spp.), prairie smoke (*Geum triflorum*), and verbenas (*Verbena* spp.).

GROWING AND PROPAGATION

Plant in rich, moist but well-drained soil in full sun or light shade. Despite its exotic appearance, Indian pink is easy to grow. Plants form neat clumps that seldom need division. Propagate from stem cuttings taken in early summer. Only tip cuttings will root. Remove the flower buds and let the cuttings sit for 4 to 6 weeks to ensure rooting.

Indian Pink

PERFECT GARDEN COMPANIONS

Sundrops (*Calylophus* spp.) are closely related to *Oenothera*, the evening primroses. The lemon or primrose yellow flowers have four wide, blunt petals that last but a day. The flowers accent the yellow-green centers of Indian pink's flowers, and the foliage creates a fine-textured foil for the bold, stiff leaves.

Half-leaf sundrops (*Calylophus serrulatus*) is a low, shrubby plant to 8 inches tall that produces a wealth of cheerful, ¾-inch, greenish yellow flowers for much of the summer. Mature clumps are mounded and open in form. Plant in average, well-drained soil or sandy to gravelly loam in full sun or light shade. Like Indian pink, plants tolerate drought but not soggy soil. Dry, often gravel or sand prairies, open woods, and plains. Zones 3 to 9.

Hartweg's sundrops (*Calylophus hartwegii*) has 1½-inch flowers and oval leaves on 4- to 8-inch stems that form low mounds. Plant in average to sandy, well-drained soil in full sun. Dry prairies, hillsides, and open woods, often in limy soils. Zones 4 to 9.

Starry campion (*Silene stellata*) has white flowers that hang from tall spikes in summer above whorls of deep green, pointed leaves. The multistemmed clumps grow 2 to 3 feet tall. Combine them with Indian pinks in rich, moist but well-drained soil in full sun or light shade. Open woods, clearings, floodplains, and meadows. Zones 4 to 8.

Stokesia laevis

Stokes' Aster

Pronunciation	STOKES-ee-uh LEE-vis
Family	Asteraceae, Aster Family
USDA Hardiness Zones	5 to 9
Native Habitat and Range	Low, wet, pine woods, bottomlands, and ditches from North Carolina to Florida and Louisiana

Stokes' Aster

DESCRIPTION

Stokes' aster produces broad rosettes of shiny green, lance-like leaves. Each leaf has a white vein that runs down the center. Branched flowerstalks rise 1 to 2 feet from the center of the clump. In summer, each stalk bears several 3- to 4-inch, flat, daisylike flowers with ragged blue petals and fuzzy white centers.

GARDEN USES

Combine Stokes' aster with perennials that have small flowers and fine-textured foliage, such as verbenas (*Verbena* spp.), penstemons (*Penstemon* spp.), yarrows, and ornamental grasses. In light shade, plant it with phlox, columbines (*Aquilegia* spp.), ferns, and hostas. In informal settings, a mass planting makes a bold statement with goldenrods, Joe-Pye weed (*Eupatorium maculatum*), sunflowers (*Helianthus* spp.), bee balm (*Monarda didyma*), blue flag iris (*Iris versicolor*), and swamp milkweed (*Asclepias incarnata*).

GROWING AND PROPAGATION

Plant in average to rich, moist soil in full sun or light shade. Stokes' asters are long-lived and easy to grow. Plants form dense clumps that are easy to divide in spring or fall. Divide for propagation or when plants get overcrowded. Lift the clumps and pull individual crowns apart. Replant even the smallest divisions; they will soon become thriving clumps. Sow ripe seeds outdoors in fall or indoors after 6 weeks of cold, moist stratification. Plants bloom the second year.

CULTIVARS OF STOKES' ASTER

'Alba' has white flowers.

'Blue Danube' has 5-inch, lavender-blue flowers.

'Blue Star' has deep sky blue flowers.

'Klaus Jelitto' has 4-inch, deep blue flowers.

'Silver Moon' has creamy white flowers.

'Wyoming' has rich blue-violet flowers.

PERFECT GARDEN COMPANIONS

Stemless ironweed (*Vernonia acaulis*) is unique among ironweeds because the narrow oval leaves are grouped in a basal rosette and the flowers are borne on bare, sparsely branched stems. The 1-inch, violet heads bloom for several weeks in midsummer. Plant in average to rich, moist soil in full sun or light shade. Zones 4 to 8.

Woodland sunflower (*Helianthus divaricatus*) has thin, dark stems with pairs of stiff, lance-shaped leaves at right angles to one another on the 2- to 6-foot stems. The 2- to 3-inch flowers bloom singly or in few-flowered clusters in mid- to late summer. Plant them at the edge of a meadow or as a backdrop to smaller plants. Plant in moist, humus-rich soil in sun or shade. Established plants tolerate dry soil. Zones 3 to 8.

Stylophorum diphyllum

Celandine Poppy

Pronunciation	sty-LOFF-or-um die-FILL-um
Family	Papaveraceae, Poppy Family
USDA Hardiness Zones	4 to 8
Native Habitat and Range	Rich, open woods, floodplains, and clearings in limy soils from Pennsylvania and Wisconsin, south to Tennessee and Arkansas

Celandine Poppy

DESCRIPTION

Celandine poppy has sea green leaves that are as decorative as its yellow-orange flowers. The leaves grow 6 to 10 inches long and resemble oak leaves. The waxy, 2-inch flowers bloom in early to midspring on 6- to 18-inch stems, covered with pairs of leaves. Plants grow from fibrous-rooted crowns.

GARDEN USES

Drifts of celandine poppies are lovely under shrubs, along woodland trails, or at the edge of a terrace garden. Combine celandine poppy with Virginia bluebells (*Mertensia virginica*), wild blue phlox (*Phlox divaricata*), Meehan's mints (*Meehania* spp.), phacelias (*Phacelia* spp.), and other blue flowers. Other suitable companions include bloodroot (*Sanguinaria canadensis*), wild gingers (*Asarum* spp.), lungworts (*Pulmonaria* spp.), epimediums (*Epimedium* spp.), merrybells (*Uvularia* spp.), sedges (*Carex* spp.), and ferns.

GROWING AND PROPAGATION

Plant in evenly moist, humus-rich soil in light to full shade. Plants tolerate slightly alkaline to moderately acid soil. After the seeds are released, cut the plants back, and they will produce lush new foliage. They may go dormant in midsummer if conditions become hot and dry. Plants reseed freely to produce plenty of seedlings for trading with gardening friends. If seedlings come up in unwanted places, they are easily moved to a suitable location.

PERFECT GARDEN COMPANIONS

Drooping trillium (*Trillium flexipes*) has creamy white flowers that droop on 1- to 4-inch stalks below the broad leaves. Plants grow 1 to 2 feet tall, usually in multistemmed clumps. Plant in evenly moist, humus-rich soil in light to full shade along with celandine poppy and other wildflowers. Zones 3 to 8.

Red trillium or **stinking Benjamin** (*Trillium erectum*) has deep blood red flowers that nod slightly on 1- to 4-inch stalks. It is called stinking Benjamin because the flowers are unpleasantly scented. Plants grow 1 to 2½ feet tall, and the lush leaves may be 10 inches across. The red flowers stand out against celandine poppy's orange blooms. Plant in evenly moist, humus-rich, acid soil in shade. Vasey's trillium (*Trillium vaseyi*) is similar, but the flowers are nodding and have wider petals. Zones 4 to 9.

Thalictrum dioicum

Early Meadow Rue

Pronunciation	thuh-LICK-trum die-OH-ih-kum
Family	Ranunculaceae, Buttercup Family
USDA Hardiness Zones	3 to 8
Native Habitat and Range	Moist, deciduous woods, clearings, and roadsides from Quebec and Manitoba, south to Alabama and Missouri

Early Meadow Rue

DESCRIPTION

Early meadow rue has separate male and female plants. Male plants produce candelabra-like clusters of fuzzy, petalless flowers with pendant golden stamens that tremble with the slightest breeze. Female flowers are insignificant. The delicate, pale gray-green leaves have rounded, lobed leaflets. Plants grow 1 to 3 feet tall and bloom in early spring.

GARDEN USES

Choose early meadow rue for the shade or rock garden, along stone walls, or in masses with shrubs. Combine it with woodland wildflowers such as hepatica, rue anemone (*Anemonella thalictroides*), alumroot (*Heuchera* spp.), bloodroot (*Sanguinaria canadensis*), and anemones (*Anemone* spp.), as well as ferns and hostas. It is lovely when planted on a bank where the delicate flowers are at eye level.

GROWING AND PROPAGATION

Plant in moist, humus-rich soil in full sun or partial shade. Early meadow rue tolerates deep shade and dry soil. If leafminers are a problem, pick off and destroy infested leaves. Spray weekly with insecticidal soap at the first sign of leafminers. Remove plant debris in fall to eliminate overwintering sites. With age, plants form dense clumps that bear many bloom stalks. Plants seldom need division but may be divided in fall for propagation.

PERFECT GARDEN COMPANIONS

Long-spurred violet (*Viola rostrata*) is a delicate, stemmed violet with 1-inch, heart-shaped leaves and lavender-blue flowers with deeper blue centers. The unique feature of the flower is the 1-inch spur that rises from the lowest petal. Violets are perfect groundcovers for planting under the taller clumps of early meadow rue. Plant in moist, humus-rich, acid soil in light to full shade. Mixed woods, clearings, and streamsides. Zones 4 to 8.

Prairie alumroot (*Heuchera richardsonii*) has rounded, toothed, furry leaves and open clusters of green flowers borne on 2- to 3-foot stalks. Plants add a bold accent to early meadow rue's fine texture. Plant in average to rich, well-drained soil in full sun or partial shade. Prairies, savannas, and open woods. Zones 3 to 8.

Wooly blue violet (*Viola sororia*) has rounded to heart-shaped, hairy leaves and deep purple-blue flowers on short stalks. The leaves may reach 10 inches after flowering. This is the most common blue violet in the eastern states. 'Freckles' has pale blue flowers flecked with purple. 'Priceana', the confederate violet, has white flowers with purple-blue centers. Plant in rich, moist soil in sun or shade. Woods, clearings, meadows, and yards. Zones 3 to 9.

Thalictrum polygamum

Tall Meadow Rue

Pronunciation	thuh-LICK-trum pahl-ih-GUH-mum
Family	Ranunculaceae, Buttercup Family
USDA Hardiness Zones	3 to 8
Native Habitat and Range	Wet meadows, floodplains, streamsides, and ditches from Labrador and Ontario, south to North Carolina and Indiana

Tall Meadow Rue

DESCRIPTION

The erect stems of tall meadow rue may reach 8 feet. They are crowned by wide, dense, branched, terminal clusters of creamy white flowers in early summer. The fine, divided leaves have narrow, lobed, gray-green leaflets.

GARDEN USES

Plant tall meadow rue at the middle or rear of the border. Its airy flowers are an effective foil for larger flowers and bold foliage. Combine it with blue flag iris (*Iris versicolor*), gayfeathers (*Liatris* spp.), garden phlox (*Phlox paniculata*), sneezeweeds (*Helenium* spp.), marsh mallows (*Hibiscus* spp.), and ornamental grasses. In meadows and at pondside, combine with wild sennas (*Cassia* spp.), blue stars (*Amsonia* spp.), swamp milkweed (*Asclepias incarnata*), Carolina bush pea (*Thermopsis villosa*), sneezeweed (*Helenium autumnale*), goldenrods, asters, and sunflowers.

GROWING AND PROPAGATION

Plant in rich, moist to wet soil in full sun or light shade. Plants are adaptable to a wide range of soils, moisture, and light. They are smaller on drier sites and may tower above the surrounding plants in rich, moist soil. Plants form multistemmed clumps over time. Divide crowns in early spring or fall. Sow fresh seeds outdoors. Self-sown seedlings will appear.

OTHER RECOMMENDED *THALICTRUM* SPECIES

Purple meadow rue (*Thalictrum dasycarpum*) is similar to tall meadow rue but is more delicate. The 4- to 6-foot stems are often purple, and the upright inflorescences have many drooping, white flowers. Plant in rich, moist soil in full sun or light shade. Wet meadows, prairies, wetland margins, shores, and floodplains. Zones 3 to 8.

Waxy meadow rue (*Thalictrum revolutum*) has delicate flowers that droop from finely branched inflorescences atop 4- to 6-foot stalks. Plant in rich, moist soil in full sun or light shade. Wet meadows and woodland margins. Zones 4 to 8.

Thermopsis villosa (also listed as *T. caroliniana*)

Carolina Lupine

Pronunciation	ther-MOP-sis vill-OH-suh
Family	Fabaceae, Pea Family
USDA Hardiness Zones	3 to 9
Native Habitat and Range	Open woods, meadows, clearings, and roadsides from North Carolina to Georgia

Carolina Lupine

DESCRIPTION

Carolina lupine is an impressive species with stiff, 3- to 5-foot stalks crowned by dense, 8- to 12-inch spikelike clusters of bright lemon yellow flowers that resemble the blossoms of peas. The alternate, compound foliage is gray-green. Each leaf consists of three oval leaflets with fuzzy undersides.

GARDEN USES

Carolina lupines have a wide spread, so leave them plenty of room in the middle or back of the border. Combine their spiky form with phlox, peonies, meadow rues (*Thalictrum* spp.), bellflowers (*Campanula* spp.), catmints (*Nepeta* spp.), and hardy cranesbills (*Geranium* spp.). Plants are also well suited to meadows and sunny wild gardens in the company of bowman's root (*Gillenia trifoliata*), bergamot (*Monarda fistulosa*), bluestars (*Amsonia* spp.), and grasses.

GROWING AND PROPAGATION

Plant in average to rich, moist but well-drained, acid soil in full sun or light shade. In warm southern zones, partial shade is recommended. Plants become anemic and turn yellow in limy soils. If foliage declines after flowering, cut plants back to the ground. Plants form tight, erect clumps that increase in breadth each year but seldom need division. Propagate by stem cuttings taken in early summer. Sow seeds outdoors or indoors after soaking them for 12 hours in hot water to break the hard seedcoat.

PERFECT GARDEN COMPANIONS

Greater coreopsis (*Coreopsis major*) bears pairs of three-lobed, stalkless leaves on 2- to 3-foot stems, giving the look of six whorled leaves. The 2-inch, yellow flowers bloom in summer in loose terminal clusters. Plant in average to rich, moist soil in full sun or partial shade. Open woods and roadsides. Zones 4 to 8.

Tall coreopsis (*Coreopsis tripteris*) has stiff, 6- to 9-foot stems clothed in 4-inch, three-lobed, lance-shaped leaves and topped with wide clusters of 2-inch, starry yellow flowers. Plant them as a backdrop for Carolina lupine and other tall plants. Plant in average to rich, moist, well-drained soil in full sun. Open woods, meadows, savannas, prairies, and bottomlands. Zones 3 to 9.

Tiarella cordifolia

Allegheny Foamflower

Pronunciation	tee-uh-REL-uh core-dih-FOE-lee-uh
Family	Saxifragaceae, Saxifrage Family
USDA Hardiness Zones	3 to 8
Native Habitat and Range	Rich, deciduous or mixed coniferous woods, streamsides, woodland coves, and rocky slopes from Nova Scotia and Wisconsin, south to Georgia and Alabama

Allegheny Foamflower

DESCRIPTION

Allegheny foamflower is a spring-blooming groundcover with erect, 6- to 10-inch conical clusters of small, fuzzy, white flowers. The heart-shaped to triangular leaves may be toothed or untoothed. Rosettes of foliage grow from fibrous-rooted crowns, and plants spread via long, leafy runners.

GARDEN USES

Allegheny foamflower and its cultivars are extraordinary groundcovers that do double duty with lovely flowers and dense, weed-excluding mats of foliage. Plant them under shrubs and trees, alone or in combination with other plants. In the shade garden, they are effective with merrybells (*Uvularia* spp.), hepatica (*Hepatica* spp.), woodland phlox (*Phlox divaricata*), Virginia bluebells (*Mertensia virginica*), wild columbine (*Aquilegia canadensis*), bloodroot (*Sanguinaria canadensis*), bleeding hearts (*Dicentra* spp.), Solomon's seals (*Polygonatum* spp.), and ferns.

GROWING AND PROPAGATION

Plant in evenly moist, humus-rich, slightly acid soil in partial to full shade. Plants spread quickly from runners to form broad, dense, leafy mats. They may swamp delicate plants as they spread. Divide plants in spring or fall; remove rooted runners anytime during the growing season. Sow seeds on top of the soil in spring.

SOME RECOMMENDED SELECTIONS OF ALLEGHENY FOAMFLOWER

'Running Tapestry' has 2½-inch leaves with three indistinct lobes. The leaf veins are deep maroon.

'Slick Rock' is a petite plant with 2-inch leaves that have five deeply cut, jagged lobes.

T. cordifolia* var. *collina (also listed as *T. wherryi*, Wherry's foamflower) produces leafy clumps with no runners and a profusion of flowerstalks. The leaves are glossy green. Many cultivars are available with various leaf forms.

'Dunvegan' is a cultivar of *T. cordifolia* var. *collina* that has pale pink flowers and deeply cut, rounded lobes that are widest above the middle.

'Oakleaf' is a cultivar of var. *collina* that has 4-inch leaves with deeply cut, rounded lobes. The center lobe is most pronounced.

Tradescantia virginiana

Virginia Spiderwort

Pronunciation	trad-es-KANT-ee-uh ver-jin-ee-AH-nuh
Family	Commelinaceae, Spiderwort Family
USDA Hardiness Zones	4 to 9
Native Habitat and Range	Moist, open woods, flood-plains, meadows, and prairies from Maine and Wisconsin, south to Georgia and Missouri

Virginia Spiderwort

DESCRIPTION

Virginia spiderwort bears plentiful blue to purple, satiny, three-petaled flowers in clusters atop slender 2- to 3-foot, succulent, jointed stems in late spring and early summer. Each flower lasts only half a day, closing by early afternoon amid the narrow, deep green, 12-inch leaves. Plants often go dormant after flowering but may re-emerge in fall.

GARDEN USES

Plant Virginia spiderwort in meadows, at the edge of woodlands, or in beds and borders. Plant it in drifts or scattered clumps with leafy plants such as everblooming bleeding heart (*Dicentra eximia*), blue stars (*Amsonia* spp.), mountain mints (*Pycnanthemum* spp.), asters, and ferns that will fill the void left when it goes dormant. In formal situations, combine Virginia spiderwort with phlox, hardy cranesbills (*Geranium* spp.), and bergenias (*Bergenia* spp.). Use it in combination with groundcovers to underplant shrubs.

GROWING AND PROPAGATION

Plant in average to rich, moist but well-drained soil in full to light shade. Plants grow well in partial shade but do not flower as long. Clumps often get tattered after flowering and are best cut to the ground. They produce new foliage quickly where summers are cool, or by autumn in warmer zones. Divide them every 2 to 3 years, as they go dormant. Self-sown seedings are often abundant.

MORE *TRADESCANTIA* SPECIES

Bracteated spiderwort (*Tradescantia bracteata*) is a stout species with blue-gray foliage and deep medium blue flowers. Plants grow 1½ feet tall with an equal spread. Plant in lean, well-drained soil in full sun or light shade. Dry prairies and savannas from Indiana and North Dakota, south to Missouri and Kansas. Zones 3 to 8.

Ohio spiderwort (*Tradescantia ohiensis*) is a slender, branching plant to 3 feet tall with narrow blue-green leaves and blue, rose, or white flowers with rounded petals. Plant in average to rich, well-drained soil in full sun or light shade. Open woods, savannas, meadows, and prairies from Massachusetts and Minnesota, south to Florida and Texas. Zones 3 to 9.

Spiderwort (*Tradescantia subaspera*) is the largest and stoutest spiderwort. Plants grow to 3½ feet tall from thick upright stems with broad, gray-green leaves and 1½-inch, blue to purple flowers. Plant in rich, moist soil in full sun or partial shade. Rich woods, clearings, and roadsides from Virginia and Illinois, south to Florida and Alabama. Zones 4 to 9.

Western spiderwort (*Tradescantia occidentalis*) grows to 2 feet tall with branched stems. Plant in lean, well-drained soil in full sun or light shade. Savannas, dry prairies, and high plains from Wisconsin and Montana, south to Texas and Utah. Zones 3 to 9.

Trillium grandiflorum

White Trillium, Large-Flowered Trillium

Pronunciation	TRIL-ee-um gran-dih-FLOOR-um
Family	Liliaceae, Lily Family
USDA Hardiness Zones	3 to 9
Native Habitat and Range	Rich, deciduous woods, rocky slopes, and floodplains from Quebec and Minnesota, south to Georgia and Indiana

White Trillium

DESCRIPTION

White trillium has broad, three-petaled, snow white flowers held erect above a whorl of bright green, broadly oval, stalkless leaves. The flowers fade to pink as they age. 'Flore-pleno' is a rare and enchantingly beautiful, fully double-flowered form.

GARDEN USES

Trilliums are a welcome addition to any garden where their simple needs can be met. Plant them in shade and woodland gardens with wildflowers, shade perennials, and ferns. They are lovely planted with spring-blooming shrubs and trees. Combine them with bloodroot (*Sanguinaria canadensis*), Virginia bluebells (*Mertensia virginica*), merrybells (*Uvularia* spp.), cranesbills (*Geranium* spp.), primroses, epimediums (*Epimedium* spp.), wild bleeding heart (*Dicentra eximia*), and sedges (*Carex* spp.).

GROWING AND PROPAGATION

Plant in moist, humus-rich soil in shade. Plants emerge early in the spring and bloom for 2 weeks. If the soil stays moist, the foliage persists through the summer. Plants may form offsets and with age can be divided. Propagation by seed is slow. Sow fresh seeds outside in summer. Germination takes a year, and plants will bloom in 5 to 7 years.

ANOTHER *TRILLIUM* SPECIES WITH STEMMED FLOWERS

Snow trillium (*Trillium nivale*) is a diminutive plant with egg-shaped, gray-green leaves and small, three-petaled, snow white flowers in earliest spring. Plants grow only 6 inches high and often go dormant after flowering. Plant in moist, humus-rich, limy soil in light to partial shade. Found in rich, rocky, deciduous woods and on outcroppings and river bluffs from Pennsylvania and Minnesota, south to West Virginia and Nebraska. Zones 4 to 8.

AVOID WILD-COLLECTED TRILLIUMS

Most trilliums sold today are collected from the wild. Buy only nursery-propagated plants from reputable dealers. Nursery-grown does not mean nursery-propagated. Check sources and ask dealers where they get their plants to avoid encouraging further decimation of wild populations.

Trillium sessile

Toad Trillium, Toadshade

Pronunciation	TRIL-ee-um SES-sill-ee
Family	Liliaceae, Lily Family
USDA Hardiness Zones	4 to 8
Native Habitat and Range	Rich, deciduous or mixed coniferous woods, rocky slopes, woodland coves, and floodplains from Pennsylvania and Illinois, south to Virginia and Arkansas

Toad Trillium

DESCRIPTION

The name toad trillium arises from the brown and green mottled leaves that resemble the back of a toad. The sweet-scented, deep maroon, 2½-inch flowers are sessile (stalkless) and stick straight up from the leaves. Plants grow up to 1 foot high and may go dormant after flowering. This species is often confused in the trade with *Trillium cuneatum*, especially when plants have been collected from the wild.

GARDEN USES

Toad trilliums are striking, but the dark flowers show up best against a backdrop of lush foliage. Combine them with wild gingers (*Asarum* spp.), creeping phlox (*Phlox stolonifera*), hepaticas (*Hepatica* spp.), merrybells (*Uvularia* spp.), and ferns.

GROWING AND PROPAGATION

Plant in moist, humus-rich soil in light to partial shade. Plants form multistemmed clumps from buds on the rhizome. Plants develop more quickly from seeds than other trilliums, and self-sown seedlings reach blooming size in 3 to 5 years. Sow ripe seeds outdoors in summer. This species is still gathered from the wild despite its ease of propagation. Buy only nursery-propagated plants.

OTHER *TRILLIUM* SPECIES WITH SESSILE (STALKLESS) FLOWERS

Prairie trillium (*Trillium recurvatum*) is smaller than *T. sessile*, with narrower leaves and 1-inch, red-brown flowers with backward-curved sepals. Plants grow ½ to 1½ feet tall. Plant in rich, moist, limy soil in light to full shade. Found in open woods, clearings, floodplains, and pine barrens from Michigan and Nebraska, south to Alabama and Mississippi. Zones 4 to 9.

Yellow trillium (*Trillium luteum*) is nearly identical to *T. sessile* except the flowers are chartreuse to lemon yellow and smell of lemons. Plant in moist, humus-rich soil in light to partial shade. Found in rich, deciduous woods and woodland coves from Virginia to Georgia and Alabama. Zones 4 to 8.

Uvularia grandiflora

Great Merrybells, Large-Flowered Bellwort

Pronunciation	you-view-LAH-ree-uh gran-dih-FLOOR-uh
Family	Liliaceae, Lily Family
USDA Hardiness Zones	3 to 8
Native Habitat and Range	Deciduous woodlands, rocky slopes, and bottomlands, usually in limy soils from Quebec to Minnesota, south to North Carolina mountains and Oklahoma

Great Merrybells (back)

DESCRIPTION

The sea green leaves of great merrybells complement its 2-inch, nodding, lemon-yellow flowers. Each bell-shaped flower consists of three showy petals and three petallike sepals. The thin, wiry 1- to 1½-foot stems pierce the blades of the 4½-inch, stalkless, oblong-oval leaves. As the plants emerge, the leaves are attractively curled, allowing a clear view of the early spring flowers. After flowering, the foliage expands to form a soothing, soft green groundcover.

GARDEN USES

My earliest memory of great merrybells is in a Virginia garden. Great clumps grew in a wooded rock garden, where the luscious, pleated leaves complemented drifts of epimediums (*Epimedium* spp.), double bloodroot (*Sanguinaria canadensis* 'Multiplex'), and a host of ferns. I use graceful merrybells as an accent in plantings of low wildflowers such as hepaticas (*Hepatica* spp.), spring beauties (*Claytonia* spp.), foamflowers (*Tiarella* spp.), creeping phlox (*Phlox stolonifera*), and wild gingers (*Asarum* spp.). They also make a lovely groundcover under shrubs in combination with ferns, alumroots (*Heuchera* spp.), sedges (*Carex* spp.), and small-leaved hostas.

GROWING AND PROPAGATION

Plant great merrybells in moist, well-drained, humus-rich to rocky, near-neutral soil in partial to full shade. In rich garden soil, plants spread quickly from creeping rhizomes to form tight clumps. Under woodland conditions, plants spread more slowly and are more open in form. Divide overgrown clumps after flowering or in the fall. Tease the individual crowns apart and replant immediately. Sow fresh seed outdoors as soon as it ripens. Germination requires cold, moist stratification followed by warm, moist stratification. Seedlings develop slowly and will bloom in 3 to 5 years.

ANOTHER RECOMMENDED *UVULARIA* SPECIES

Perfoliate bellwort (*Uvularia perfoliata*) is similar to great merrybells but is smaller and more delicate. Stems may reach 1½ feet tall. The flowers are 1 inch long and pale greenish yellow. Plant in dry to moist, humus-rich, acid to near-neutral soil in partial to full shade. Plants grow naturally in deciduous woods from Quebec and New England, south to Florida and Louisiana. Zones 4 to 8.

Veratrum viride

False Hellebore, American White Hellebore

Pronunciation	ver-AT-rum VEER-ih-dee
Family	Liliaceae, Lily Family
USDA Hardiness Zones	2 to 8
Native Habitat and Range	Low, wet woods, streamsides, seeps, bogs, and wet meadows from Quebec and Alaska, south to North Carolina, Montana, and Oregon

DESCRIPTION

The 12-inch, pleated oval leaves of false hellebore overlap one another on thick, 1- to 6-foot stalks. The terminal flower clusters have drooping branches covered in tightly packed, starry green flowers in early summer. Each flower consists of three narrow petals and three petallike sepals. Plants go dormant by midsummer unless they bear ripening seed. The fading foliage turns rich yellow.

False Hellebore

GARDEN USES

Bold foliage and unusual flowers make false hellebore a unique garden plant. Use it in bog gardens or in wet woods and along streams. Combine it with marsh marigold (*Caltha palustris*), skunk cabbage (*Symplocarpus foetidus*), cardinal flower (*Lobelia cardinalis*), purple avens (*Geum rivale*), tassel rue (*Trautvetteria carolinensis*), and ferns. The emerging spring stalks look like pleated spikes and contrast well with ephemeral wildflowers such as spring beauties (*Claytonia* spp.), wood anemone (*Anemone quinquefolia*), and trout lilies (*Erythronium* spp.).

GROWING AND PROPAGATION

Plant false hellebore in moist to wet, humus-rich soil in full sun or full shade. Plants bloom sparingly where shade is dense. Flowering is sporadic even among mature plants. A clump may bloom one year and not the next. Mature clumps have huge, deep crowns that are difficult to move. Sow fresh seed outdoors as soon as it is ripe. Seedlings develop slowly.

A MORE PETITE *VERATRUM* SPECIES

Small-flowered veratrum (*Veratrum parviflorum*) is a diminutive species by comparison to *V. viride*, with three to five broad, oval basal leaves and tall, leafless stalks to 3 feet with ½-inch, lime green flowers. Plant in moist, humus-rich, acid soil in light to full shade. Found in rich woods, on rocky slopes, and near springs in the mountains from Virginia, south to North Carolina and Tennessee. Zones 4 to 7.

Verbena canadensis

Rose Verbena

Pronunciation	ver-BEAN-uh kan-uh-DEN-sis
Family	Verbenaceae, Vervain Family
USDA Hardiness Zones	5 to 10
Native Habitat and Range	Open woods, clearings, waste places, and roadsides from Pennsylvania and Illinois, south to Florida and Texas

Rose Verbena

DESCRIPTION

Rose verbena has 8- to 18-inch, trailing, wiry stems with sharply toothed and lobed, oval, opposite leaves. The flat terminal flower clusters elongate into short spikes as the showy purple, rose, or white flowers open. The ¼- to ¾-inch flowers are tubular with flat, five-petaled faces. They bloom throughout the summer and into fall.

GARDEN USES

Rose verbenas are excellent plants to use to tie a mixed planting together. Let them creep among foliage plants such as artemisias (*Artemisia* spp.), bergenias (*Bergenia* spp.), mulleins (*Verbascum* spp.), and ornamental grasses, or use them to bridge gaps between flowering perennials such as purple coneflowers (*Echinacea* spp.), yarrows, black-eyed Susans (*Rudbeckia* spp.), and phlox. In meadows and seaside gardens, plant them with blanket flowers (*Gaillardia* spp.), yuccas (*Yucca* spp.), butterfly weed (*Asclepias tuberosa*), gayfeathers (*Liatris* spp.), asters, and goldenrods. In rock gardens, let them creep among dwarf conifers, or use them to fill the spaces left when bulbs go dormant.

GROWING AND PROPAGATION

Plant rose verbenas in well-drained, sandy or loamy soil in full sun to light shade. Verbenas are tough, heat- and drought-tolerant perennials that bloom tirelessly during the summer. They spread quickly, rooting at the leaf joints as they trail along the ground, to form showy groundcovers. When flowering becomes sparse, shear plants back to encourage fresh growth and flowers. They are easily grown from stem cuttings taken anytime during the growing season. Seeds need 3 to 4 weeks of cold, moist stratification to germinate. Seedlings develop quickly.

OTHER RECOMMENDED *VERBENA* SPECIES

Blue verbena (*Verbena hastata*) is a graceful, erect verbena with branched, candelabra-like spikes of small blue flowers in summer and long, narrow, toothed leaves. Plants grow 3 to 5 feet tall and form multistemmed clumps. Plant in rich, evenly moist to wet soil in full sun or light shade. Combine them with bee balms (*Monarda* spp.), phlox, and grasses to add a strong vertical accent to borders and beds. Plants grow naturally in wet meadows and prairies, wetland margins, marshes, and ditches from Nova Scotia and British Columbia, south to Florida and Arizona. Zones 3 to 8.

Hoary vervain (*Verbena stricta*) is an upright verbena with hairy, wedge-shaped leaves on branched 1- to 1½-foot stems. In summer, the terminal spikes bear blue-violet flowers. Plant in average, sandy or loamy, well-drained soil in full sun or light shade. Plants grow naturally in dry prairies, savannas, clearings, and waste places and along roadsides from Ontario and Wyoming, south to Texas and New Mexico. Zones 3 to 8.

Vernonia noveboracensis

New York Ironweed

Pronunciation	ver-NO-nee-uh no-vee-bore-uh-SEN-sis
Family	Asteraceae, Aster Family
USDA Hardiness Zones	4 to 9
Native Habitat and Range	Low meadows, floodplains, pond margins, and springs from Massachusetts and Ohio, south to Florida and Mississippi

New York Ironweed

DESCRIPTION

The 1-inch, red-violet flowers of New York ironweed are carried in broad, showy clusters on leafy, 4- to 6-foot stems from late summer into fall. The 6- to 8-inch leaves are lance-shaped and mostly toothless. The plants presumably get the name ironweed from the prominent, rust-red hairs in the spent flower heads and on the fruits.

GARDEN USES

Ironweeds are commanding border plants. Place them toward the back of the border, and give them ample room to grow to mature size. Plant them with yarrows, cannas (*Canna* spp.), lilies, salvias, and irises. They are well suited to meadow and prairie gardens as well as to pond banks. Combine them with hibiscus, phlox, great blue lobelia (*Lobelia siphilitica*), Joe-Pye weeds (*Eupatorium* spp.), turtleheads (*Chelone* spp.), asters, goldenrods, and grasses.

GROWING AND PROPAGATION

Plant in rich, evenly moist soil in full sun or light shade. Plants are easy to grow and thrive under most garden situations. Leafminers may make pale tunnels in the leaves. Prune off and destroy infested leaves. Spray leaves weekly with insecticidal soap at the first sign of leafminers. Clumps of New York ironweed get very large in time but seldom need division. If plants are too crowded, thin out one quarter of the stems in late spring. Take stem cuttings in early summer. Self-sown seedlings will appear.

OTHER RECOMMENDED *VERNONIA* SPECIES

Narrowleaf ironweed (*Vernonia augustifolia*) has a different growth habit from all other *Vernonia* species. The 1-inch, deep violet flowers bloom in broad, open clusters on wide, shrublike clumps from 2 to 4 feet tall. The 6- to 8-inch leaves are narrowly lance-shaped and droop slightly on the stems. Plant in average to rich, evenly moist soil in full sun or light shade. Found in sandy meadows, open woods, and savannas and along roadsides from North Carolina, south to Florida and Mississippi. Zones 4 to 9.

Tall ironweed (*Vernonia altissima*) is the giant of the group, with 4- to 10-foot stems and dense, broad, flattened clusters of soft violet flowers in late summer and autumn. Plant in rich, moist soil in full sun. Plants grow naturally in wet meadows and prairies, wetland edges, pond margins, and ditches from New York, Michigan, and Nebraska, south to Georgia and Louisiana. Zones 4 to 8.

Veronicastrum virginicum

Culver's Root

Pronunciation	ver-on-ih-KAS-trum ver-JIN-ih-kum
Family	Scrophulariaceae, Figwort Family
USDA Hardiness Zones	3 to 9
Native Habitat and Range	Open woods, low meadows, prairies, floodplains, and outcroppings from Ontario and Manitoba, south to Georgia and Louisiana

Culver's Root (white)

DESCRIPTION

From midsummer to early fall, the erect, creamy white candelabra spikes of Culver's root make a dramatic show on thick, 3- to 6-foot stems. The toothed, 2- to 6-inch, lance-shaped leaves are borne in tiered whorls on the stems. The variety *rosea*, sometimes listed as 'Roseum', has pale rose-pink flowers. Soft blue selections are seen but are likely hybrids of the European species.

GARDEN USES

The upright spikes of Culver's root add lift and excitement to the middle or rear of beds and borders. Combine it with phlox, turtleheads (*Chelone* spp.), hibiscus, gauras (*Gaura* spp.), purple coneflowers (*Echinacea* spp.), and yarrows. In meadows and prairies, plant it with milkweeds (*Asclepias* spp.), bee balms (*Monarda* spp.), ironweeds (*Vernonia* spp.), anise hyssop (*Agastache foeniculum*), rattlesnake master (*Eryngium yuccifolium*), asters, goldenrods, and grasses.

GROWING AND PROPAGATION

Plant Culver's root in rich, moist soil in full sun or light shade. Plants tolerate some dryness but wilt with prolonged drought. Plants form multistemmed clumps that are easily divided in early spring or fall. Sow seeds outdoors in fall or indoors in spring with 4 weeks of cold, moist stratification.

PERFECT GARDEN COMPANIONS

Aromatic aster (*Aster oblongifolius*) produces mounds of 1¼-inch, purple flowers that cover the 16-inch stems in midautumn. A real showstopper, this outstanding plant blooms for at least 3 weeks in September and early October. The fuzzy, oblong foliage and scaly buds create an interesting display through the summer season with taller plants such as Culver's root. Give plants average, well-drained soil in full sun or light shade. They flop miserably in too much shade or rich soils. 'Dream of Beauty' is shorter in stature and has rose-pink flowers. 'Raydon's Favorite' is a southern cultivar with blue-purple flowers. Found on dry slopes and in prairies, open woods, and savannas. Zones 3 to 8.

Lead plant (*Amorpha canescens*) is a loose shrub that grows to 4 feet tall with twice-compound leaves sporting ¾-inch leaflets clothed in soft, gray hairs. The foliage is attractive throughout the season. The 6-inch bloom spikes bear tightly packed, tubular, purple flowers with conspicuous, protruding orange stamens. Lead plant's mounded form is a nice complement to the upright spikes of Culver's root. In structured beds and borders, cut plants back to the ground every year or two to keep the growth from looking twiggy. Established plants are impossible to move successfully. Give plants average to rich, moist but well-drained soil in full sun or light shade. Established plants tolerate drought well. Found in loamy or sandy, moist to dry open woods, savannas, and prairies. Zones 3 to 8.

Viola canadensis

Canada Violet

Pronunciation	vy-OH-luh kan-uh-DEN-sis
Family	Violaceae, Violet Family
USDA Hardiness Zones	3 to 8
Native Habitat and Range	Deciduous or mixed coniferous woods, rocky slopes, and floodplains from Newfoundland and Alberta, south to South Carolina and Arizona

Canada Violet

DESCRIPTION

Canada violet is a stemmed violet to 1 foot tall with broad, heart-shaped leaves and white flowers with a purple blush on the back of the petals. Each five-petaled flower has two petals that point upward and three that point outward and down. Plants bloom for up to 3 weeks in spring. They spread by underground runners to form dense patches. Striped violet (*Viola striata*) is often confused with Canada violet, but its creamy white flowers have narrower petals that lack the purple blush on the back.

GARDEN USES

Canada violets form attractive groundcovers under shrubs and flowering trees. In informal gardens, plant them with bulbs, hostas, and early-blooming perennials. They make good companions for wildflowers such as trilliums, baneberries (*Actaea* spp.), Virginia bluebells (*Mertensia virginica*), merrybells (*Uvularia* spp.), bloodroot (*Sanguinaria canadensis*), and ferns.

GROWING AND PROPAGATION

Plant in moist, humus-rich soil in sun or shade. Canada violets grow under a wide range of soil and moisture conditions. In rich soil, plants are robust, heavy bloomers. In dry sites, plants flower for a short time and go dormant in summer. Plants spread readily enough to be considered somewhat invasive. Divide plants after flowering or in fall if you wish to increase their numbers.

OTHER STEMMED *VIOLA* SPECIES

Dog violet (*Viola conspersa*) is a delicate plant with 1-inch leaves and 1½-foot stems bearing small, medium blue flowers with short spurs. The small, heart-shaped leaves are dark green. Plant in rich, moist soil in sun or shade. Found in rich, open woods, clearings, floodplains, and streamsides from Quebec and Minnesota, south to South Carolina and Alabama. Zones 3 to 8.

Downy yellow violet (*Viola pubescens*) is a robust violet with soft, hairy leaves and lemon yellow flowers. Plants grow 6 to 16 inches tall and form dense clumps but do not spread by runners. Plant in rich, evenly moist soil in light to full shade. Found in deciduous and mixed woods, floodplains, and clearings from Nova Scotia and North Dakota, south to Georgia and Oklahoma. Zones 3 to 8.

Viola pedata

Birdfoot Violet, Pansy Violet

Pronunciation	vy-OH-luh pe-DAH-tuh
Family	Violaceae, Violet Family
USDA Hardiness Zones	3 to 9
Native Habitat and Range	Dry meadows and prairies, savannas, and rocky or gravely embankments from Maine and Minnesota, south to Florida and Texas

Birdfoot Violet

DESCRIPTION

This diminutive, stemless violet has deeply lobed leaves that resemble the foot of some fanciful bird. The flat-faced, powder blue or bicolor blue and purple flowers are carried on 4- to 6-inch stalks in early spring.

GARDEN USES

Birdfoot violet is best used in rock gardens or on dry banks where it will not be overwhelmed by more exuberant plants. Combine it with campions (*Silene* spp.), prairie smoke (*Geum triflorum*), blue-eyed grass (*Sisyrinchium* spp.), rue anemone (*Anemonella thalictroides*), and grasses. Try it in troughs and containers with diminutive plants such as bluebells (*Campanula* spp.), saxifrages, and pinks (*Dianthus* spp.). Plants will form large, self-sown colonies on sandy embankments and along gravel driveways.

GROWING AND PROPAGATION

Plant in average, sandy or loamy, well-drained soil in full sun or light shade. Plants may be difficult to establish and will rot if the soil is too wet and heavy. In good sites, they form large clumps smothered in flowers. Birdfoot violets may be short-lived, but self-sown seedlings are plentiful under ideal conditions. Established plants rebloom sporadically in summer and fall. Divide clumps after flowering in summer. Sow seed outdoors when it ripens.

OTHER STEMLESS *VIOLA* SPECIES

Larkspur violet (*Viola pedatifida*) is similar to *V. pedata* but is much larger, with lobed and dissected leaves to 8 inches tall after flowering. The flower is curved instead of flattened and is borne on a 2- to 3-inch stem before the leaves expand. Plants may rebloom in autumn. Plant in average to rich, well-drained soil in full sun or light shade. Found on moist and dry prairies and savannas and along roadsides. Zones 3 to 9.

Wild blue violet (*Viola cucullata*) has wavy, heart-shaped leaves and 1-inch, purple-blue flowers in spring. Plant in moist to wet, humus-rich soil in light to full shade. Plants form dense, multicrowned clumps that are easily divided after flowering or in autumn. Found in wet woods, streamsides, seeps, and bogs. Zones 3 to 8.

Viola sororia

Woolly Blue Violet

Pronunciation	vie-OH-luh so-ROAR-e-ah
Family	Violaceae, Violet Family
USDA Hardiness Zones	3 to 9
Native Habitat and Range	Woods, clearings, meadows, and yards from Quebec and Minnesota south to North Carolina and Oklahoma

Woolly Blue Violet

DESCRIPTION

Woolly blue violet is the species most people picture when they think of violets. It is a stemless species with rounded to heart-shaped hairy leaves and deep purple-blue flowers on short stalks. The leaves may reach 10 inches tall after flowering. This is the most common blue violet in the eastern states.

GARDEN USES

Woolly blue violets form a lush, green groundcover where few other plants will grow. Plant them in ordinary woodland soil with a host of wildflowers such as rue anemone (*Anemonella thalictroides*), bellworts (*Uvularia* spp.), mayapple (*Podophyllum peltatum*), and baneberry (*Actaea pachypoda*), as well as sedges and ferns. They grow well in beds and borders with perennials, bulbs, and ornamental grasses.

GROWING AND PROPAGATION

Plant woolly blue violets in average to humus-rich, moist soil in sun or shade. They tolerate a wide variety of conditions and are the most common violet found in woods and gardens. Plants form dense, multi-crowned clumps that are easily divided after flowering or in autumn. In summer and autumn, plants produce unusual, petalless flowers that never open, called cleistogamous flowers. They are self-pollinated and produce a bumper crop of seeds. As a result, self-sown seedlings are plentiful. In fact, these violets spread so well, they often make a nuisance of themselves. Spider mites occasionally attack the foliage in dry weather.

CULTIVARS OF WOOLLY BLUE VIOLET

'Freckles' has pale blue flowers flecked with purple. **'Priceana'**, the confederate violet, has white flowers with purple-blue centers.

ANOTHER GARDEN-WORTHY VIOLET

Primrose-leaved violet (*Viola primulifolia*) is a stemless violet with deep green, elongated arrowhead-shaped leaves to 10 inches tall. The sweet-scented, small white flowers are streaked in the center with purple. The flowers are borne before the leaves. After blooming, the plants form dense clumps from slow-creeping rhizomes. Plant in rich, moist to wet soil in light to full shade. Plants also grow in wet, sandy soils. Found in low woods, seeps, streambanks, and wet meadows from New Brunswick and Wisconsin, south to Florida and Texas. Zones 4 to 9.

Waldsteinia fragarioides

Barren Strawberry

Pronunciation	wald-STEIN-ee-uh fray-gah-ree-OY-deez
Family	Rosaceae, Rose Family
USDA Hardiness Zones	4 to 8
Native Habitat and Range	Open woods, savannas, clearings, embankments, and outcroppings from Maine and Minnesota, south to Pennsylvania and Indiana, and in the mountains to North Carolina

Barren Strawberry

DESCRIPTION

The deep green leaves of barren strawberry have three rounded, toothed lobes and are carried in open rosettes. The ½-inch, bright yellow flowers are borne in late spring. The flowers resemble those of true strawberries, except their petals are yellow instead of white. Unlike edible strawberries, they do not produce succulent fruits.

GARDEN USES

Use barren strawberry as a groundcover in the dry shade of shrubs and trees, on difficult slopes, and in rock gardens. Combine it with milkweeds (*Asclepias* spp.), yarrows, blanket flowers (*Gaillardia* spp.), bowman's-root (*Gillenia trifoliata*), baptisias (*Baptisia* spp.), campions (*Silene* spp.), phlox, and grasses. Use them to edge beds and paths or plant them between pavers in walkways to create a soft, lush effect.

GROWING AND PROPAGATION

Plant in average, sandy or loamy soil in full sun to partial shade. Plants are drought- and heat-tolerant. They form broad colonies from creeping stems that root at the nodes; colonies are easily divided in spring or fall.

AN EDIBLE *WALDSTEINIA* RELATIVE

Wild strawberry (*Fragaria virginiana*) is a delicate plant with three-lobed, toothed, hairy leaves and white flowers in spring. The edible red berries ripen in early summer. The plants spread to form an open groundcover. Plant in average to rich, well-drained soil in full sun to light shade. Found in open woods, savannas, meadows, and prairies and along roadsides throughout most of North America. Zones 2 to 9.

A PERFECT GARDEN COMPANION

Mouse-ear coreopsis (*Coreopsis auriculata*) is a low, spreading groundcover with fuzzy, 2- to 5-inch triangular leaves and 2-inch, yellow-orange flowers held 1 to 2 feet above the foliage in spring. Creeping stems advance the clumps steadily outward, but plants are seldom invasive. Interplant the two groundcovers for an interesting textural carpet with blooms in both spring and summer. Mouse-ear coreopsis grows well in partial or open shade and retains its attractive foliage all season in moist soil. 'Nana' is a compact cultivar. Found in open woods and clearings. Zones 4 to 9.

Yucca filamentosa

Adam's-Needle, Yucca

Pronunciation	YUK-uh fill-uh-men-TOE-suh
Family	Agavaceae, Agave Family
USDA Hardiness Zones	4 to 10
Native Habitat and Range	Sand dunes, outcroppings, and pine barrens, mainly along the coastal plain and in the mountains from Maryland to Georgia

DESCRIPTION

Adam's-needle brings the desert to mind with its rosettes of stiff, blue-green, 2- to 2½-foot, sword-shaped leaves. In summer, erect, multibranched bloom stalks rise 5 to 15 feet above the foliage, bearing nodding, creamy white flowers. Each waxy flower has three petals and three petallike sepals that form a 2-inch-long bell. 'Bright Edge' has yellow-variegated leaves.

GARDEN USES

Use Adam's-needle as an accent plant in dry soil, rock gardens, or seaside plantings. Contrast its stiff foliage with soft or delicate plants such as sedums, gauras (*Gaura* spp.), blanket flowers (*Gaillardia* spp.), evening primroses (*Oenothera* spp.), sages (*Salvia* spp.), and verbenas (*Verbena* spp.). In formal designs, use the stiff, swordlike foliage to mark the corners of beds or to flank a sidewalk or entryway. Plants are beautiful in pots and decorative urns underplanted with creeping plants such as swan river daisy (*Brachycome* spp.), lantana, helichrysum, and verbena. They naturalize in thin or rocky soils where few other plants thrive.

GROWING AND PROPAGATION

Plant Adam's-needle in average to rich, well-drained soil in full sun or light shade. The plants will thrive for years with little care. The main crown dies after flowering, but auxiliary crowns keep the plants growing. Divide by removing young side shoots from the clump in spring or fall.

Adam's-Needle

OTHER RECOMMENDED *YUCCA* SPECIES

Soapweed (*Yucca glauca*) has stiff, narrow, gray-green leaves to 2 feet long and dense clusters of 2½- to 3-inch flowers. Plant in average to rich, well-drained soil in full sun or light shade. Plants grow naturally on dry prairies and plains. Zones 3 to 9.

Spanish bayonet (*Yucca aloifolia*) has stiff, 2½-foot, deep green to bluish green leaves in a dense cluster atop thick trunks. Plants grow in dense clumps with many trunks, some as tall as 10 feet. The flowers are carried in open clusters, and the seedpods droop on the stems. Plant in dry or moist, sandy soil in full sun or light shade. Plants grow naturally on sand dunes and at the edges of tidal marshes. Zones 7 to 11.

Weakleaf yucca (*Yucca flaccida*) is similar to *Y. filamentosa* but its leaves are thinner and droop at the tips. Plant in average, sandy or loamy soil in full sun or light shade. Plants grow naturally in dry, open woods, pine barrens, and dunes. Zones 4 to 9.

Zizia aptera

Heart-Leaved Alexanders

Pronunciation	ZIZ-ee-uh AP-ter-uh
Family	Apiaceae, Parsley Family
USDA Hardiness Zones	3 to 9
Native Habitat and Range	Moist to seasonally wet prairies, low meadows, clearings, and open woods from New York to British Columbia, south to Georgia and Nevada

Heart-Leaved Alexanders

DESCRIPTION

Heart-leaved Alexanders has basal rosettes of deep green, shiny, heart-shaped, 4- to 6-inch leaves. The stem leaves consist of three to five toothed, oval leaflets. Flattened heads of yellow flowers provide a bright accent in the spring garden, where yellow is a welcome color. The flower clusters are 1 to 1½ inches across. The lush foliage is attractive all summer and turns shades of wine in autumn. Plants grow 1 to 2 feet tall with an open form. This parsley relative is a larval food plant for many butterflies, including swallowtails.

GARDEN USES

In prairies and meadows, place heart-leaved Alexanders next to the path so that they can be appreciated at close range. Prairie smoke (*Geum triflorum*), phlox, mountain mints (*Pycnanthemum* spp.), alumroots (*Heuchera* spp.), and milkweeds (*Asclepias* spp.) are good companions. Combine them with columbines (*Aquilegia* spp.), wild blue phlox (*Phlox divaricata*), and cranesbills (*Geranium* spp.) along woodland walks. The open, mounded form of heart-leaved Alexanders makes it a good plant for linking other perennials in the front or middle of the border. Plant them there amid cranesbills (*Geranium* spp.), catmints (*Nepeta* spp.), yuccas (*Yucca* spp.), and grasses.

GROWING AND PROPAGATION

Plant in average to humus-rich, moist soil in full sun to moderate shade. Established plants are drought-tolerant. They form full-size clumps in a few years' time but seldom require division. Sow seeds indoors in winter with cold, moist stratification. Self-sown seedlings will appear. Divide plants in autumn.

ANOTHER RECOMMENDED *ZIZIA* SPECIES

Golden Alexanders (*Zizia aurea*) is a bushy plant with many leafy stems. The leaves are finely divided and are more delicate than those of *Z. aptera*. The flower clusters are larger, up to 2½ inches across. Plant in rich, moist or dry soil in full sun or partial shade. Plants grow naturally in open woods, floodplains, meadows, and prairies from Quebec and Saskatchewan, south to Florida and Texas. Zones 3 to 9.

Part III

Designing with Wildflowers

◀ *A lush meadow garden of colorful black-eyed Susan, bee balm, bergamot, and purple coneflower surround a tranquil lawn. An enchanting woodland trail begins at the vine-covered arbor.*

Designing with Wildflowers

You may not think of garden design when it comes to wildflowers—just throw the seeds down and you'll have a meadow, right? But as you'll see from the gardens featured here, you can create beautiful wildflower gardens for every yard and situation once you stop thinking of wildflowers as "meadows in a can" and start thinking of them as garden-worthy perennials. The gardens in this section make the most of wildflowers' many colors, shapes, and textures. They feature wildflowers in every part of the yard: along a driveway, surrounding a small lawn, under trees, around a terrace, and in a sunny border. The designs range from casual to formal. And they use wildflowers to solve landscape problems, such as what grows well in dry shade, deep shade, or a hot sunny site; what to plant around a terrace; and what to grow as a lovely alternative to a lawn.

Bear in mind that these designs are just jumping-off points. They combine different wildflower gardens in three yard settings to inspire you to try wildflowers in your own yard. Combine any of the gardens shown here to suit your tastes. Or pick the part of a design that you like best—even if it's just a great plant combination—and try that.

When creating your own designs with wildflowers, use the same principles you would when designing a perennial garden. Use plants that bloom at different times to create a long flowering season. Mix heights, shapes, and textures to give the garden depth and dimension—if everything looks the same, you'll quickly be bored by it. Make sure you include plants with attractive and colorful foliage, such as ferns, ornamental grasses, and heucheras, to keep the garden looking interesting even when there's not a lot in bloom. Think of your garden as a tapestry where plants blend together to create a beautiful effect. And get ready to enjoy the visiting birds, butterflies, toads, and other wild friends that will share your pleasure.

This section features three complete garden designs. Each of the designs is divided into two smaller gardens to make it easier to adapt the designs to your yard. This gives you the flexibility to use one garden alone or to mix and match the gardens to meet your needs. For each of the three garden designs, there is an introduction, followed by descriptions of the two smaller gardens. Next comes the plant lists keyed to the plans of the gardens.

This colorful front yard prairie garden features blazing-stars, bergamot, oxeyes, black-eyed Susans, and little bluestem.

I've gotten together with two other garden designers, Fred Rozumalski and Edith Eddleman, to create wildflower gardens for many styles and situations. The first design is a front yard garden, since I like to make the point that wildflowers are showy enough to shine in the front yard as well as the back. It features two gardens, Fred's prairie garden, which forms colorful borders around a central lawn, and my savanna garden, which divides into two small beds that are great for tight, sunny spaces.

The second design covers a situation that's common to many yards, including mine—some areas are in full sun, while others are shaded. So in this design, a garden for sun and shade, you'll find Edith's sunny formal garden and my shaded terrace garden. The third design celebrates the country feel that wildflowers can give your property. I've included two favorite types of wildflower gardens, a woodland wildflower garden and a wildflower meadow garden. So turn to the designs, picture them in your own yard, pick your favorites, and enjoy!

A Front Yard Garden

This front yard garden has two distinct areas: a sunny prairie garden and a savanna garden in partial shade. The sidewalk forms a natural division between the two areas. The garden is in a cottage style to complement a charming Victorian house, but its formal lines and the exuberance of the planting make it a welcome addition to any yard. Although the entire front yard is made into a garden here, you can adapt any section of the garden design to fit a smaller site. For example, you can use the bed in front of the house alone by ending it at the edge of the porch and along the stone pathway. Or copy the section along the street sidewalk and use it as a freestanding bed or border. It's easy to divide the savanna garden into two smaller beds along the central path.

A Prairie Garden

Designed by Fred Rozumalski

For some people, the mention of a prairie or meadow conjures up images of wildness. This misperception of the sometimes fuzzy look of prairie plants is counterbalanced in this garden by the orderly structure of the design. Here rows of grasses and wildflowers along the sidewalk create a living fence for the front yard that also forms an interior space. The "fence" turns the formerly exposed, unused front yard into a comfortable place to be. The continuous display of color and the formal planting scheme provide a feeling of neatness that your neighbors will appreciate.

This prairie garden contains a large number of wildflowers to ensure constant bloom and to attract butterflies and bees. Prairie smoke ushers in spring, followed quickly by golden Alexanders, blue-eyed grass, prairie coral bells, and prairie phlox. Summer brings waves of color from the yellows of prairie coneflowers, oxeyes, and goldenrods, to the white spikes of Culver's root, the pink heads of Joe-Pye weed, and the purple spires of gayfeathers. Leave the spent flowers and seedheads on the plants to feed birds during the long winter and to provide winter interest. Golden dried grasses standing in the clean snow provide an extra spark of interest. As in a natural prairie, the grasses are the unifying element in this garden. The repeated vertical structure of the stems gracefully ties the great diversity of wildflowers together.

Fred Rozumalski is a horticulturist and landscape architect practicing ecosystem restoration and sustainable landscape design in Minneapolis/St. Paul, Minnesota. He specializes in storm water conservation and wetland restoration.

A Savanna Garden

Designed by C. Colston Burrell

Dry, semishaded areas are often the hardest to keep attractive with traditional garden plants—they just can't compete with aggressive tree roots. Not so with native plants. But these tough wildflowers are up to the challenge. Summer dryness is seldom a problem; these plants are naturally drought-tolerant. If conditions get really dry, plants adapt to extreme stress by going dormant rather than dying. They will disappear for the season but will reappear unharmed next spring.

Bright morning sun floods this planting, but by midday, the sun is filtered by the open canopy of the trees. By afternoon, the trees and the house together cast the garden into full shade.

In this garden, flowers and foliage combine to produce a pleasing season-long display. The shade trees are the focal point of the garden with a backdrop provided by the front of the house. Spring bloom comes early with the nodding rose puffs of prairie smoke and the tassels of early meadow rue. Columbines, spiderworts, wild geraniums, and golden Alexanders soon create a bright combination in midspring. For late spring and early summer, a variety of penstemons lift their spikes of tubular flowers among the lustrous heart-shaped leaves of the Alexanders and the blue-gray foliage of starry Solomon's plume. High summer belongs to flowering spurge, which weaves its small, white flowers through the foliage and stems of other plants. Late-flowering asters and goldenrods provide autumn color in concert with the yellow and burgundy of fall foliage, while little bluestem grasses offer both autumn and winter interest. Their silky plumes reflect sunlight, and the graceful stems sway with every breeze.

Plants for the Prairie Garden

The number to the left of each plant name indicates the plant's location(s) in the design.
The number to the right shows how many of each plant the design includes.

Wildflowers

1. Lead plant (*Amorpha canescens*)—8
2. Butterfly weed (*Asclepias tuberosa*)—16
3. Smooth aster (*Aster laevis*)—6
4. New England aster (*Aster novae-angliae*)—2
5. Upland white aster (*Aster ptarmicoides*)—25
6. Purple prairie clover (*Dalea purpurea*)—28
7. Joe-Pye weed (*Eupatorium purpureum*)—3
8. Prairie smoke (*Geum triflorum*)—53
9. Oxeye (*Heliopsis helianthoides*)—4
10. Prairie coral bells (*Heuchera richardsonii*)—7
11. Rough gayfeather (*Liatris aspera*)—8
12. Kansas gayfeather (*Liatris pycnostachya*)—12
13. Wild bergamot (*Monarda fistulosa*)—2
14. Foxglove penstemon (*Penstemon digitalis*)—10
15. Prairie phlox (*Phlox pilosa*)—15
16. Virginia mountain mint (*Pycnanthemum virginianum*)—9
17. Gray-headed coneflower (*Ratibida pinnata*)—9
18. Rosinweed (*Silphium integrifolium*)—7
19. Prairie blue-eyed grass (*Sisyrinchium campestre*)—12
20. Gray goldenrod (*Solidago nemoralis*)—15
21. Stiff goldenrod (*Solidago rigida*)—2
22. New York ironweed (*Vernonia noveboracensis*)—2
23. Culver's root (*Veronicastrum virginicum*)—2
24. Golden Alexander (*Zizia aurea*)—22

Grasses

25. Switch grass (*Panicum virgatum*)—27
26. Little bluestem (*Schizachyrium scoparium*)—18
27. Indian grass (*Sorghastrum nutans*)—11
28. Prairie dropseed (*Sporobolus heterolepis*)—16

Plants for the Savanna Garden

The number to the left of each plant name indicates the plant's location(s) in the design.
The number to the right shows how many of each plant the design includes.
The "S" befrore the numbers on the left lets you know that these plants are in the savanna garden.

Wildflowers

S1. Prairie onion (*Allium stellatum*)—2
S2. Wild columbine (*Aquilegia canadensis*)—9
S3. Calico aster (*Aster lateriflorus*)—10
S4. Bigleaf aster (*Aster macrophyllus*)—6
S5. Azure aster (*Aster oolentangiensis*)—2
S6. Upland white aster (*Aster ptarmicoides*)—5
S7. Wood mint (*Blephilia ciliata*)—6
S8. Flowering spurge (*Euphorbia corollata*)—6
S9. Wild geranium (*Geranium maculatum*)—8
S10. Prairie smoke (*Geum triflorum*)—13
S11. Prairie coral bells (*Heuchera richardsonii*)—7
S12. Slender penstemon (*Penstemon gracilis*)—6
S13. Prairie phlox (*Phlox pilosa*)—3
S14. Starry Solomon's plume (*Smilacina stellata*)—8
S15. Gray goldenrod (*Solidago nemoralis*)—2
S16. Showy goldenrod (*Solidago speciosa*)—2
S17. Early meadow rue (*Thalictrum dioicum*)—5
S18. Virginia spiderwort (*Tradescantia virginiana*)—6
S19. Heart-leaved Alexander (*Zizia aptera*)—6

Ferns and Grasses

S20. Interrupted fern (*Osmunda claytoniana*)—1
S21. Little bluestem (*Schizachyrium scoparium*)—3
S22. Prairie dropseed (*Sporobolus heterolepis*)—3

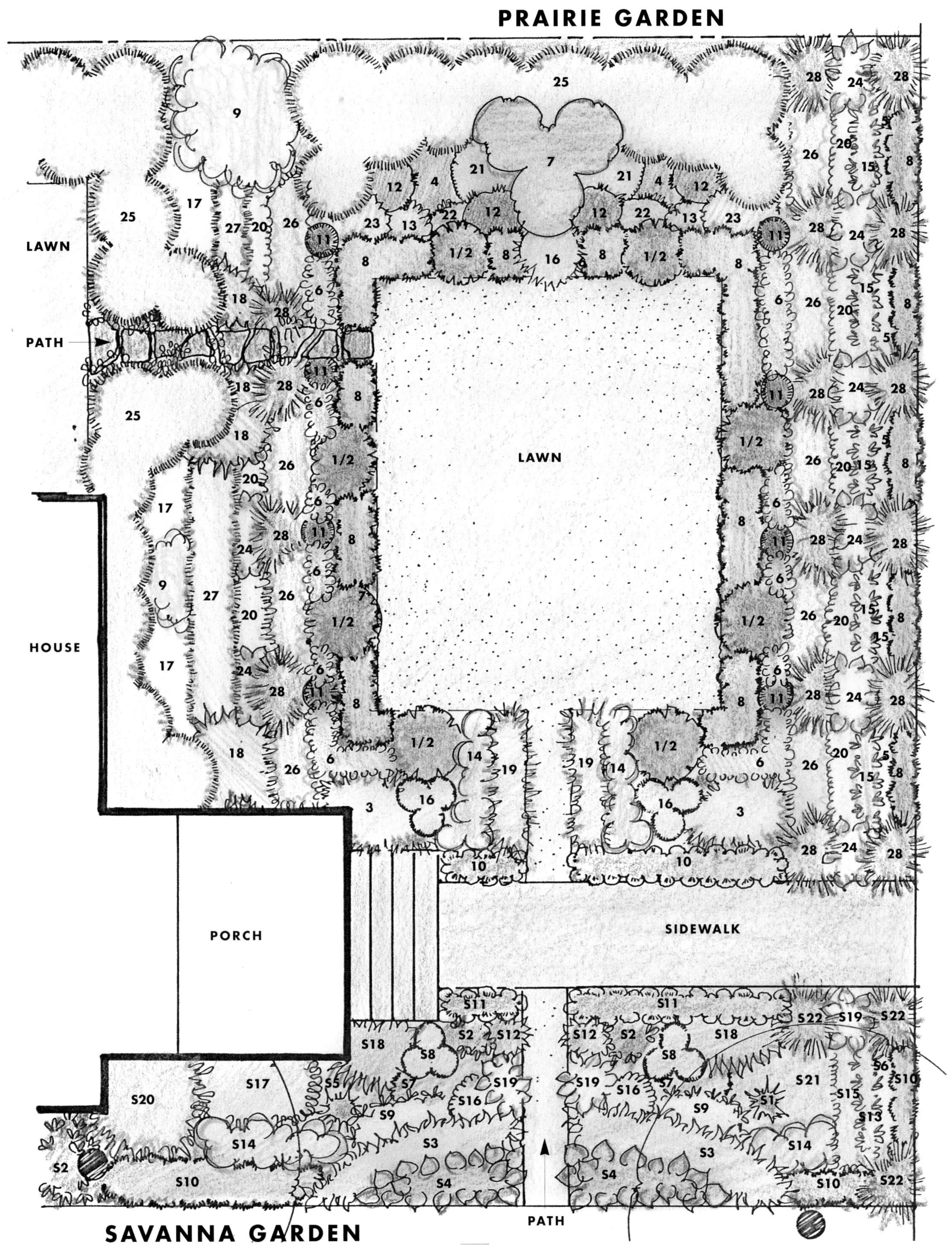
PRAIRIE GARDEN
LAWN
PATH
HOUSE
PORCH
LAWN
SIDEWALK
PATH
SAVANNA GARDEN

A Garden for Sun and Shade

This spacious garden turns a long, narrow yard into a garden paradise. The formal sunny garden is filled to overflowing with lush plantings. The tranquil shade garden becomes a cool retreat from summer's heat, as a restful tapestry of soothing green leaves follows the lush carpet of spring flowers. A row of viburnums separates the sunny garden from the shaded terrace, providing privacy in the spring before the front garden grows tall. It's easy to change both the sunny formal garden and the shaded terrace garden to fit smaller spaces. Convert the long sides of the sunny garden into smaller beds or borders, or use the stepping stone path and the hedge to divide the garden into smaller plantings. The other sections of the garden may also be divided to fit smaller spaces. You can use either side of the terrace garden, from the hedge to the stone path, as a separate bed against a house or around the front of a shaded deck.

A Sunny Formal Garden

Designed by Edith R. Eddleman

This sunny garden provides a formal setting for plants native throughout the eastern, central, and southern United States. Plants were selected for their beauty, toughness, color, and length of bloom. A sea of plants encloses a rectangular lawn, creating an island of tranquility in the midst of a busy neighborhood. Cool, misty colors and see-through plants such as camassia, verbena, and veronica have been used because to our eyes, they look farther away than they are, making the beds appear deeper. Additional layers of color and texture increase the luxurious feeling of the planting.

Yuccas, liatris, tall larkspurs, penstemons, skullcaps, and grasses provide upright counterpoints to the more rounded forms of purple coneflowers, bee balms, phlox, and asters. The towering giants Joe-Pye weed, scarlet rose mallow, and the tallest grasses lift the viewer's eye toward the sky.

A lengthy season of bloom keeps this garden colorful through fall. Spring's lavender-blue phlox, red-violet penstemons, pink and mauve verbenas, lilac irises, and yellow sundrops give way to summer's pink and white phlox, purple coneflowers, and misty blue delphiniums and skullcaps. Cotton-candy pink clouds of queen-of-the-prairie seem to be served up on the dessert-plate–size hibiscus blooms, which are soon followed by the dusty mauve domes of Joe-Pye weed.

Autumn brings another flush of verbena and penstemon blooms, joined by violet-blue asters, yellow silkgrass, pink turtleheads, and coppery red grasses as well as the huge scarlet flowers of scarlet rose mallow. In winter, the stiff evergreen rosettes of yuccas give a quiet distinction to the garden's walks and lawns.

This large garden can be scaled down for smaller lots by dividing it into sections. Each of the three main sections of the garden, the front and the two sides, easily stands alone as a delightful smaller garden.

Edith R. Eddleman is a garden designer, lecturer, and photographer living in Durham, North Carolina. She specializes in perennial garden design and enjoys designing for public gardens. She is a volunteer curator and designer of the perennial borders at the North Carolina State University Arboretum.

Purple coneflowers brighten sunny spots in summer with perennials such as common yarrow.

A Shaded Terrace Garden

Designed by C. Colston Burrell

A terrace in the light shade of spring is an ideal place for shaking off the last of winter's chill. Colorful, fragrant wildflowers are a welcome antidote for winter's blues. This garden features a planting of woodland wildflowers, ferns, and sedges in moist soil under a shade tree. In nature, all the flowers growing in this garden flourish in deciduous woodlands in the shade of maples, oaks, and basswoods.

Privacy for the outdoor seating area is provided by a mixed planting of native shrubs. The shrubs were chosen for their flowers as well as for their fruits, which are an important food source for birds. They also add year-round structure and summer nest sites. A small clay basin of still water creates a focal point and allows birds to drink and bathe. An old stump from a fallen tree makes an enchanting planter for ferns.

Bloodroot has bright white spring flowers and lush foliage in summer.

Dwarf crested iris makes a delicate accent along the edge of a terrace or path.

The design features not only the spring floral display but also an intricate textural carpet of foliage throughout the summer. Earliest spring brings the furry flowers of hepaticas with their delicate, sweet scent. A host of woodland wildflowers such as Dutchman's breeches, spring beauties, trout lilies, and false rue anemones soon add to the display. The burst of color is glorious but short-lived. These ephemeral wildflowers are interplanted with later- blooming species that will fill the spaces left when the early-blooming plants go dormant. The fragrant Virginia bluebells and wild sweet William are always showstoppers. To fill the gaps left when the early wildflowers go dormant, wild ginger, bellworts, creeping phlox, and ferns take center stage as the last flowers of wild geranium give way to the white spires of black snakeroot. The bold foliage of spikenard adds a tropical touch throughout the summer. The summer berries of Solomon's plume and spikenard are relished by cardinals, thrushes, and other birds. An autumn display is provided by white wood asters, red baneberries, and a host of goldenrods. Red chokeberry and arrowwood are laden with berries throughout fall and winter. Evergreen ferns and sedges take the garden through the late season until snow creates an insulating blanket. The dried seedheads of black cohosh punctuate the snow, adding winter interest and providing seeds for birds.

Plants for the Sunny Formal Garden

The number to the left of each plant name indicates the plant's location(s) in the design.
The number to the right shows how many of each plant the design includes.

Wildflowers

1. Nodding onion (*Allium cernuum*)—5
2. Prairie onion (*Allium stellatum*)—16
3. Thimbleweed (*Anemone cylindrica*)—5
4. Swamp milkweed (*Asclepias incarnata*)—3
5. Heart-leaved aster (*Aster cordifolius*)—8
6. New England aster (*Aster novae-angliae*)—2
7. Aromatic aster (*Aster oblongifolius*)—8
8. Blue false indigo (*Baptisia minor*)—1
9. Half-leaf sundrops (*Calylophus serrulatus*)—5
10. Pink turtlehead (*Chelone lyonii*)—3
11. Green and gold (*Chrysogonum virginianum*)—6
12. Narrowleaf silkgrass (*Chrysopsis graminifolia*)—3
13. Tall larkspur (*Delphinium exaltatum*)—8
14. Purple coneflower (*Echinacea purpurea*)—11
15. Joe-Pye weed (*Eupatorium fistulosum*)—3
16. Flowering spurge (*Euphorbia corollata*)—10
17. Queen-of-the-prairie (*Filipendula rubra*)—5
18. Prairie smoke (*Geum triflorum*)—3
19. Scarlet rose mallow (*Hibiscus coccineus*)—2
20. Rose mallow (*Hibiscus moscheutos*)—2
21. Southern blue flag (*Iris virginica* 'Contraband Girl')—3
22. Rocky Mountain blazing-star (*Liatris ligulistylis*)—5
23. Kansas gayfeather (*Liatris pycnostachya*)—5
24. Dotted horsemint (*Monarda punctata*)—5
25. Bee balm (*Monarda didyma*)—2
26. Sundrops (*Oenothera fruticosa*)—3
27. Foxglove penstemon (*Penstemon digitalis*)—3
28. Small's beardtongue (*Penstemon smallii*)—19
29. Moss phlox (*Phlox subulata*)—17
30. Garden phlox (*Phlox paniculata*)—6
31. Virginia mountain mint (*Pycnanthemum virginianum*)—2
32. Downy skullcap (*Scutellaria incana*)—3
33. Rose verbena (*Verbena canadensis*)—20
34. Hoary vervain (*Verbena stricta*)—8
35. Narrowleaf ironweed (*Vernonia angustifolia*)—1
36. Culver's root (*Veronicastrum virginicum*)—3
37. Yucca (*Yucca filamentosa*)—5
38. Weakleaf yucca (*Yucca flaccida*)—6

Grasses

39. Bent awn plume grass (*Erianthus contortus*)—1
40. Sugarcane plume grass (*Erianthus giganteus*)—1
41. Little bluestem (*Schizachyrium scoparium* 'The Blues')—3

Plants for the Shaded Terrace Garden

The number to the left of each plant name indicates the plant's location(s) in the design.
The number to the right shows how many of each plant the design includes.
The "T" befrore the numbers on the left tells you that these plants are in the shaded terrace garden.

Wildflowers

T1. Red baneberry (*Actaea rubra*)—3
T2. Spikenard (*Aralia racemosa*)—1
T3. Canada wild ginger (*Asarum canadense*)—10
T4. White wood aster (*Aster divaricatus*)—9
T5. Black snakeroot (*Cimicifuga racemosa*)—5
T6. Spring beauty (*Claytonia virginica*)—12 (interplanted among white wood asters)
T7. Dutchman's breeches (*Dicentra cucullaria*)—12 (interplanted among foamflowers)
T8. Shooting star (*Dodecatheon meadia*)—12
T9. Yellow trout lily (*Erythronium americanum*)—12 (interplanted among white wood asters)
T10. Wild geranium (*Geranium maculatum*)—7
T11. Sharp-lobed hepatica (*Hepatica acutiloba*)—12
T12. Crested iris (*Iris cristata*)—12
T13. False rue anemone (*Isopyrum biternatum*)—12 (interplanted among blue phlox)
T14. Virginia bluebells (*Mertensia virginica*)—12 (interplanted among great merrybells)
T15. Wild sweet William (*Phlox divaricata*)—10
T16. Bloodroot (*Sanguinaria canadensis*)—9
T17. Solomon's plume (*Smilacina racemosa*)—11
T18. Wreath goldenrod (*Solidago caesia*)—3
T19. Zigzag goldenrod (*Solidago flexicaulis*)—6
T20. Allegheny foamflower (*Tiarella cordifolia*)—11
T21. Great merrybells (*Uvularia grandifolia*)—6

Ferns

T22. Maidenhair fern (*Adiantum pedatum*)—17
T23. Lady fern (*Athyrium filix-femina*)—7
T24. Leather wood fern (*Dryopteris marginalis*)—8
T25. Ostrich fern (*Matteuccia struthiopteris*)—22
T26. Narrow beech fern (*Thelypteris phegopteris*, also listed as *Phegopteris connectilis*)—2
T27. Christmas fern (*Polystichum acrostichoides*)—3

Shrubs

T28. Red chokeberry (*Aronia arbutifolia*)—2
T29. Arrowwood viburnum (*Viburnum dentatum*)—4

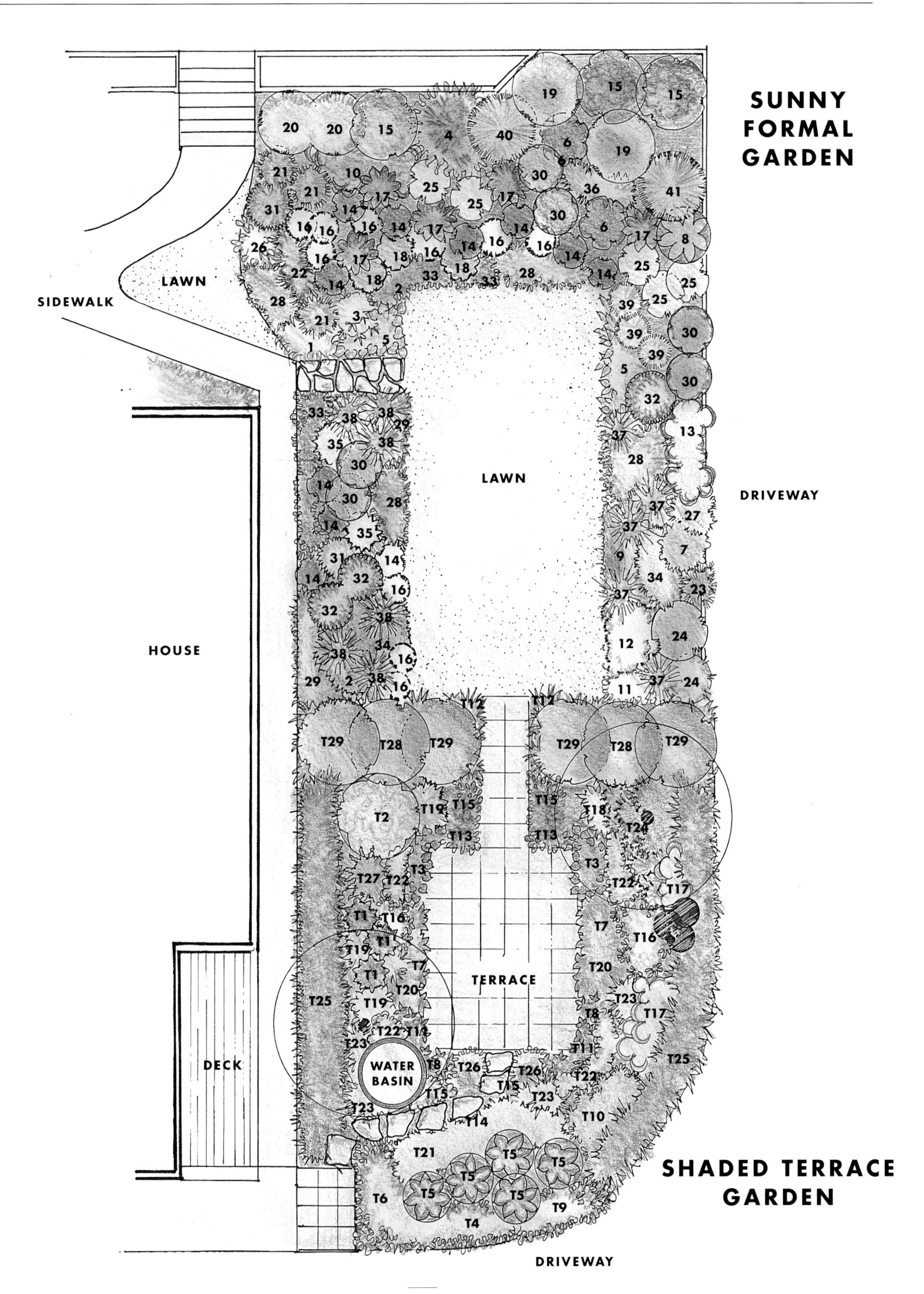
SUNNY
FORMAL
GARDEN
SIDEWALK
LAWN
LAWN
HOUSE
DRIVEWAY
TERRACE
DECK
WATER
BASIN
SHADED TERRACE
GARDEN
DRIVEWAY

A Country Garden

This large garden is suitable for open spaces and a sweeping view. It suits the open feel of the country landscape and brings back the woodland and meadow wildflowers that were once familiar sights along country lanes. The garden has a two sections, a shaded woodland grove and a colorful meadow. The gardens meet at the narrowest point in the yard, where the woods nearly reach the lawn.

The shade garden contains many interesting plant combinations that you can use in smaller gardens. Ferns mix with carpets of wildflowers such as phlox, wild ginger, and foamflower. These plants are perfect companions in shaded spots where grass is hard to grow. Any combination or section of the garden can be picked up from this design and adapted to a smaller space. Pick sections where smaller plants are to the front and larger plants form a backdrop for the planting and have fun mixing and matching.

The meadow garden is most effective in a larger space. Plant all or part of it in an area next to a woodland, in front of a house or at the back of a yard. The mowed grass path is a logical dividing point if you want to reduce the size of the garden to fit a smaller space.

A Woodland Wildflower Garden

Designed by C. Colston Burrell

If you're lucky enough to have an existing patch of woodland on your property (or even a cluster of large shade trees), you'll be able to enjoy the wonderful spring display of woodland flowers. In a mature woods, conditions are ideal for woodland plants, and little preparation is needed. This garden is under a canopy of maple and oak trees next to an open meadow. Feel free to substitute any shade trees. A path leads from the parking area to a small clearing in the trees where a bench could be placed. Beyond the clearing, the path enters the meadow. Fruiting viburnums and other shrubs are used to screen the parking area from the garden and to blend the planting into the remaining woods.

This garden is planted in successive waves of color that circle the clearing in a flowing pattern. Waves of spring color are produced by foamflowers, Virginia bluebells, great merrybells, phlox, and bloodroot. Wild geraniums, columbines, astilbes, and black snakeroot carry spring into early summer. In summer, a carpet of ferns, wild gingers, irises, and spurges creates a beautiful pattern, punctuated by taller plants. Fall color is nearly as showy as spring with the fruits of white baneberry, blue cohosh, Solomon's plume, and Solomon's seal. Goldenrods and asters create sprays of golden yellow and white, while the fall foliage of the ferns takes on russet hues.

A Wildflower Meadow Garden

Designed by C. Colston Burrell

As you leave the cool shade of the trees, the sun-filled clearing features a wildflower meadow garden edging the lawn. As the floral display of the woodland fades in late spring, the meadow flowers are starting to put on a show. A narrow mowed path leads through the meadow and returns to the house through the perennial garden.

Large drifts of sun-loving species are planted among native grasses. As in a prairie, the grasses create the backdrop for the flowers while helping to keep weeds down. (Where a grass plant is growing, a weed can't grow.) Unlike lawn grass, meadow grasses grow in clumps, so the flowers can grow between them. The four different grass species are mixed together and planted in a grid throughout the meadow site. Each grass is spaced two feet apart, and the wildflowers are planted within the grid as specified in the design.

Early summer brings clusters of orange butterfly weed blooms to complement penstemons, mountain mint, senna, and anise hyssop. The midsummer show comes from lavender bergamot and purple blazing-stars, coneflowers, and Joe-Pye weed. In autumn, the grasses turn russet and dry in the wind while the flowers of asters and goldenrods lure butterflies to the meadow.

The meadow garden is fairly easy to maintain once it is established. Just give it an annual spring mowing to eliminate woody plants that would eventually turn your meadow into a forest.

Pink-flowered Joe-Pye weed and cardinal flowers add bold, colorful accents in a meadow garden along the edge of a wood.

Plants for the Woodland Wildflower Garden

The number to the left of each plant name indicates the plant's location(s) in the design.
The number to the right shows how many of each plant the design includes.

Wildflowers

1. White baneberry (*Actaea pachypoda*)—8
2. Wild columbine (*Aquilegia canadensis*)—5
3. Spikenard (*Aralia racemosa*)—3
4. Canada wild ginger (*Asarum canadense*)—25
5. White wood aster (*Aster divaricatus*)—6
6. False goat's beard (*Astilbe biternata*)—4
7. Blue cohosh (*Caulophyllum thalictroides*)—6
8. Black snakeroot (*Cimicifuga racemosa*)—12
9. White wild bleeding heart (*Dicentra eximia* 'Alba')—4
10. Shooting star (*Dodecatheon meadia*)—12
11. White wild geranium (*Geranium maculatum* 'Alba')—11
12. Crested iris (*Iris cristata*)—4
13. Virginia bluebells (*Mertensia virginica*)—3
14. Allegheny pachysandra (*Pachysandra procumbens*)—2
15. Creeping phlox (*Phlox stolonifera* 'Blue Ridge')—8
16. Creeping Jacob's ladder (*Polemonium reptans*)—5
17. Great Solomon's seal (*Polygonatum biflorum* var. *commutatum*)—19
18. Bloodroot (*Sanguinaria canadensis*)—13
19. Solomon's plume (*Smilacina racemosa*)—3
20. Wreath goldenrod (*Solidago caesia*)—6
21. Allegheny foamflower (*Tiarella cordifolia*)—30
22. Great merrybells (*Uvularia grandifolia*)—5

Ferns

23. Maidenhair fern (*Adiantum pedatum*)—33
24. Lady fern (*Athyrium filix-femina*)—27
25. Interrupted fern (*Osmunda claytoniana*)—11
26. Christmas fern (*Polystichum acrostichoides*)—9

Shrubs

27. Mapleleaf viburnum (*Viburnum acerifolium*)—13
28. Arrowwood viburnum (*Viburnum dentatum*)—13

Plants for the Wildflower Meadow Garden

The number to the left of each plant name indicates the plant's location(s) in the design.
The number to the right shows how many of each plant the design includes.
The "M" befrore the numbers on the left shows that these plants are in the meadow garden.

Wildflowers

M1. Willow amsonia (*Amsonia tabernaemontana*)—7
M2. Butterfly weed (*Asclepias tuberosa*)—8
M3. Calico aster (*Aster lateriflorus*)—8
M4. New England aster (*Aster novae-angliae*)—11
M5. Azure aster (*Aster oolentangiensis*)—4
M6. Blue false indigo (*Baptisia australis*)—4
M7. Pale purple coneflower (*Echinacea pallida*)—10
M8. Joe-Pye weed (*Eupatorium fistulosum*)—5
M9. Flowering spurge (*Euphorbia corollata*)—4
M10. Bowman's root (*Gillenia trifoliata*)—6
M11. Purple-headed sneezeweed (*Helenium flexuosum*)—5
M12. Maximilian sunflower (*Helianthus maximiliani*)—3
M13. Scaly blazing-star (*Liatris squarrosa*)—9
M14. Foxglove penstemon (*Penstemon digitalis*)—14
M15. Hairy beardtongue (*Penstemon hirsutus*)—5
M16. Wild petunia (*Ruellia humilis*)—8
M17. Rosinweed (*Silphium integrifolium*)—7
M18. Early goldenrod (*Solidago juncea*)—11
M19. Rough-stemmed goldenrod (*Solidago rugosa*)—5
M20. New York ironweed (*Vernonia noveboracensis*)—1
M21. Culver's root (*Veronicastrum virginicum*)—8

Grasses

Mix these four grasses through the meadow garden. Plant them 2 feet apart, and plant the wildflowers between them.

Broom sedge (*Schyzachyrium scoparium*)—12
Little bluestem (*Schizachyrium scoparium*)—12
Indian grass (*Sorghastrum nutans*)—12
Purple-top (*Tridens flavus*)—12

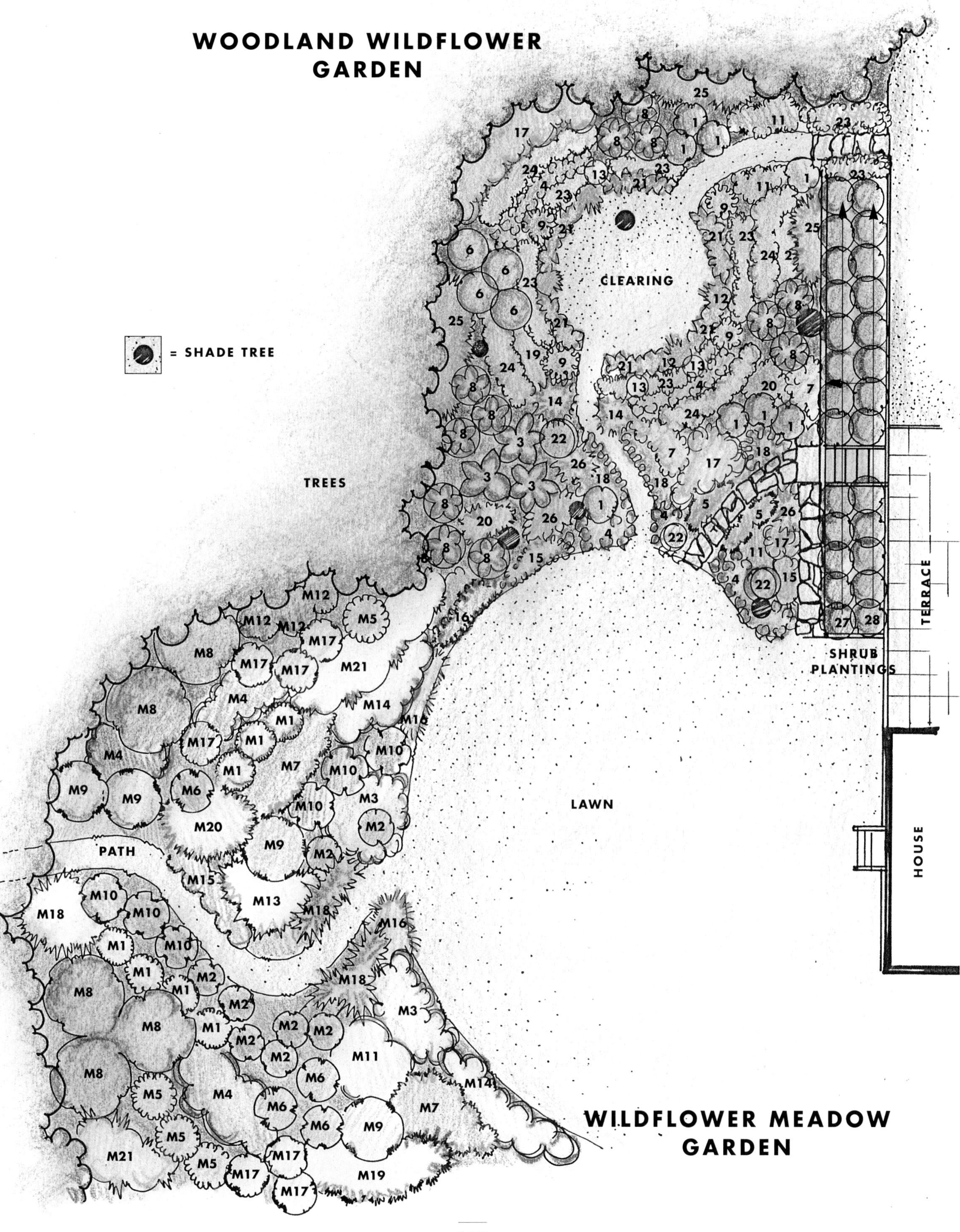
WOODLAND WILDFLOWER GARDEN
= SHADE TREE
TREES
CLEARING
LAWN
PATH
TERRACE
SHRUB PLANTINGS
HOUSE
WILDFLOWER MEADOW GARDEN

Glossary

Acid soil: Soil with a pH value less than 7.0; also referred to as sour soil.

Alkaline soil: Soil with a pH value greater than 7.0; also referred to as sweet soil.

Annual: A plant such as blanket flower (*Gaillardia pulchella*) that completes its entire life cycle in one growing season.

Axil: The area above the junction of a petiole or leaf blade with a stem.

Basal leaf: A leaf that arises directly in or from the crown of a plant.

Berry: A fleshy fruit like that of wintergreen formed from an ovary that contains many seeds.

Biennial: A plant such as fern-leaved phacelia (*Phacelia bipinnatifida*) that completes its life cycle in two seasons.

Blade: The flattened portion of a leaf.

Boreal: Referring to the northernmost extent of the continents.

Boreal forest: A northern coniferous forest characterized by a short growing season and cold, acid, highly organic soil.

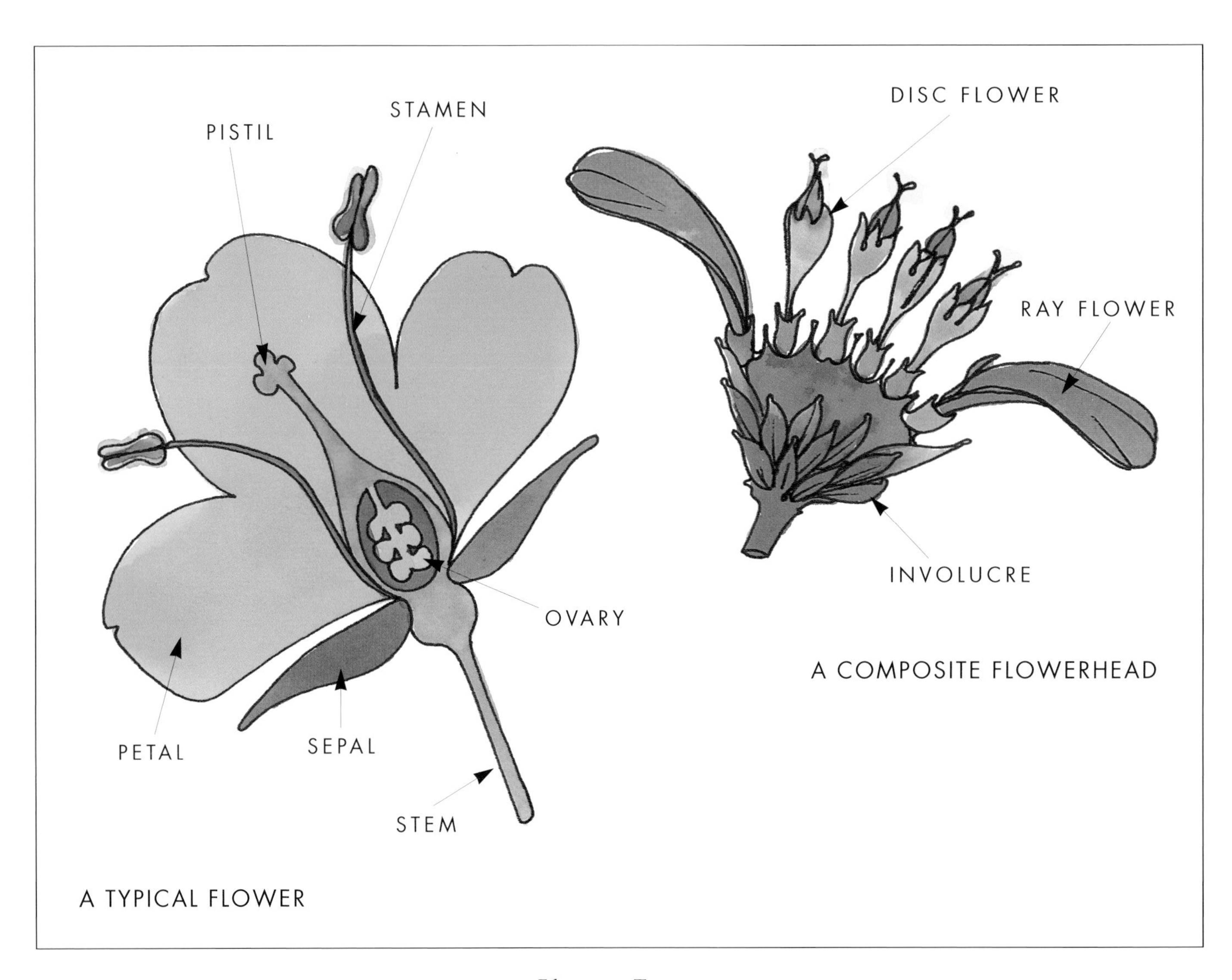

Flower Types

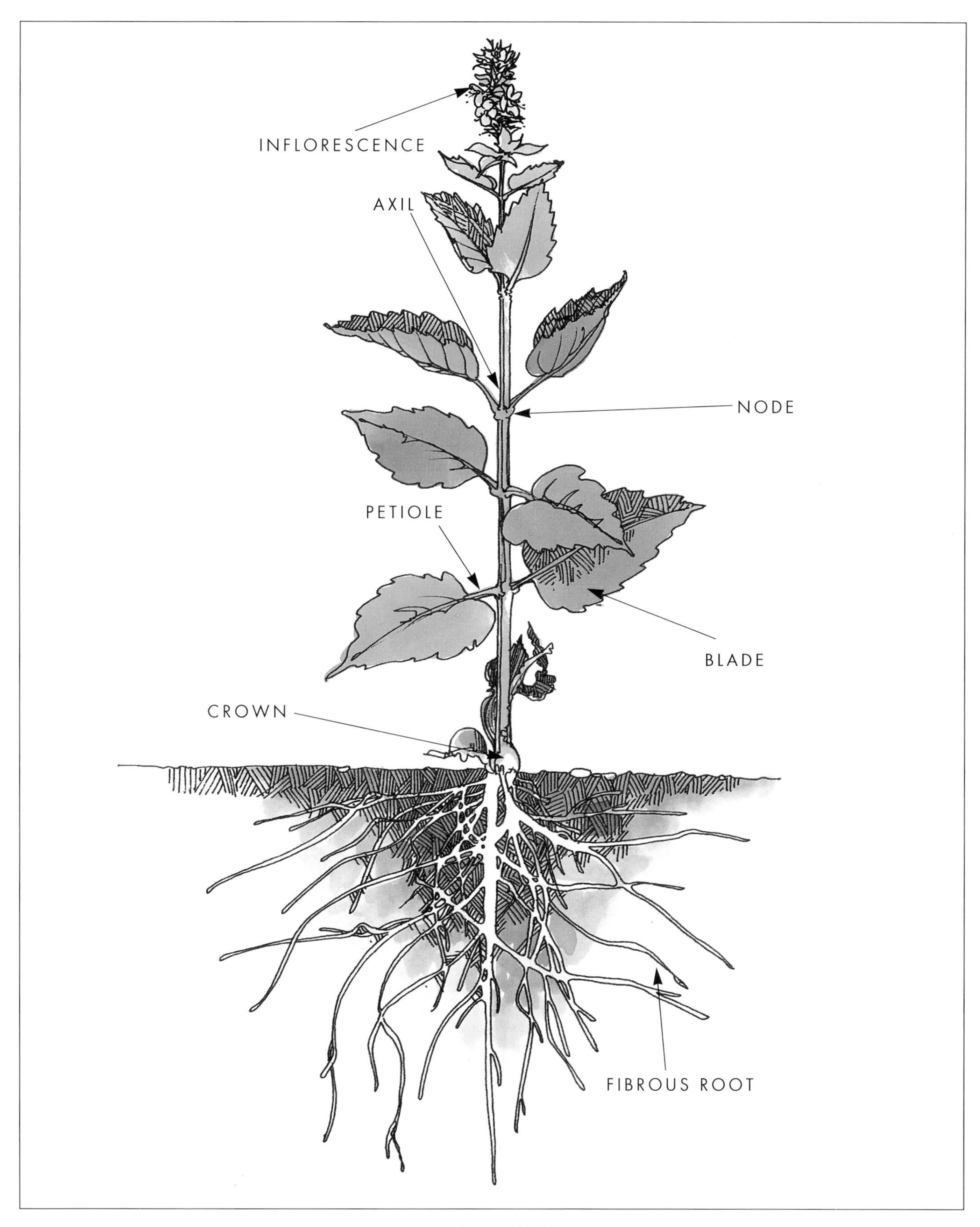

Parts of a Wildflower

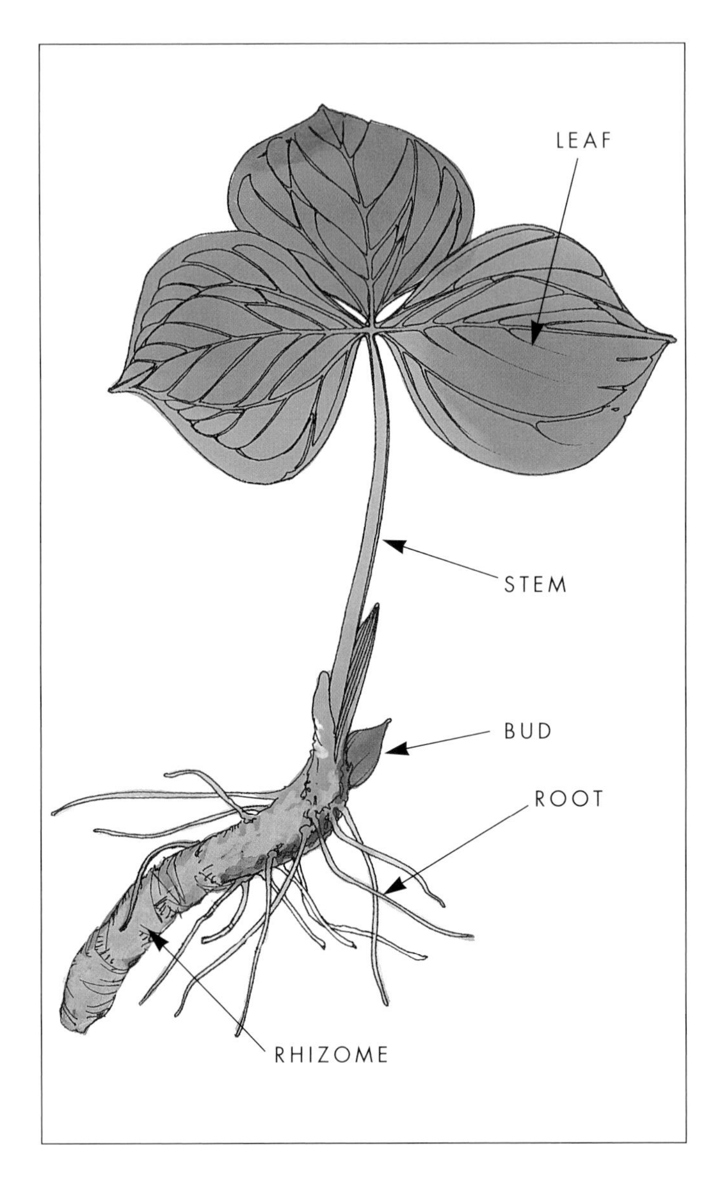

Parts of a Trillium Plant

Bract: A modified leaf usually growing near or among flowers and often colored like a petal. Bee balm "flowers" consist of both true flowers and bracts.

Bud: A small, knoblike outgrowth on the stem that may become a leaf, flower, or shoot.

Bulb: A stem consisting of thickened leaves that store food.

Capsule: A dry, multicelled fruit that splits open to release the seeds when it is ripe. Lily and iris fruits are capsules.

Clone: A plant that is derived from vegetative propagation techniques such as cuttings or division and is genetically identical to its parent.

Corm: A swollen stem base that is modified to store food; resembles a bulb.

Cove: A cool, north- or east-facing area in a wooded region, with a flattened or gradually sloping bottom of moist, very rich woodland soil that is sheltered from hot afternoon sun by trees and mountain ridges.

Crown: The part of the plant where the stem or rosette joins the roots, usually just below the soil.

Cultivar: A cultivated variety or hybrid of a wild plant that breeders have selected because it has an outstanding horticultural trait such as large flowers or variegated foliage. Cultivars are given a descriptive name, such as `Blue Ridge' creeping phlox, and are usually propagated asexually to maintain the desired trait.

Disc flowers: Small, tubular flowers found at the center of a composite flowerhead such as a purple coneflower.

Dormancy: A period of inactivity of a plant or a seed.

Ephemeral: A woodland wildflower such as spring beauty or Dutchman's breeches that grows, blooms, and sets seed in spring before becoming dormant.

Exotic plant: A plant from outside a region, state, or country, growing there by human introduction. For example, Queen-Anne's-lace is an exotic plant that now grows wild in most of the United States.

Fibrous root: A typical root growing from a crown. It absorbs water and nutrients and anchors the plant.

Forb: An herbaceous flowering plant, such as a wildflower.

Fruit: A ripened ovary.

Grass: A plant in the grass family, Poaceae, with narrow leaves on jointed stems.

Glade: A meadowlike opening in a normally forested region that remains open because it has thin soil that remains dry and cannot support growth of shrubs or trees.

Herbaceous: Nonwoody.

Inflorescence: A group of flowers.

Involucre: The receptacle that holds a composite flower.

Native plant: A plant indigenous to a particular region, state, or country. In North America, native plants are considered to be those growing prior to European settlement.

Native wildflower: An herbaceous flowering plant indigenous to a particular region, state, or country. (*See* Native plant.)

Neutral soil: Soil with a pH value of 7.0. Neutral soil is neither acidic nor alkaline.

Node: The point of attachment of a leaf to a stem.

Ovary: The portion of the pistil that contains seeds and will form a fruit.

Perennial: A plant that lives for three or more seasons and flowers more than once; this term often refers to herbaceous plants.

Petal: One of the showy, often colorful parts of a flower.

Petiole: A leaf stalk.

pH: A measure of the acidity or alkalinity of soil based on the chemical analysis of the hydrogen ion concentration of the soil.

Pistil: The female reproductive structure of a flower.

Pod: Any dry fruit such as a milkweed follicle that splits to release its seeds.

Ray flowers: The showy, petallike flowers that surround the center of a composite flower like a daisy or that make up the entire flowerhead such as in an aster.

Rhizome: An underground, horizontal stem that stores food and water; it can produce roots and shoots at its nodes.

Rosette: A basal cluster of leaves that arise from a crown.

Runner (stolon): A thin, creeping stem that forms above or below ground and sprouts roots at its nodes and new leaves and buds at its end.

Sedge: A grasslike plant in the sedge family, Cyperaceae, with unjointed, often triangular stems.

Seed: A fertilized egg (ovule) that grows into a new plant.

Sepal: One part of a whorl of green, leafy structures, located on the flower stem just below the petals.

Sessile: Lacking a leaf or fruit stem; attached directly to a stem.

Stamen: The pollen-producing part of a flower.

Stratification: A cold or warm, moist, or combination of treatments that help a seed overcome dormancy

Taproot: A thick, unbranched or sparsely branched root that stores food and water.

Tuber: A short, fleshy, underground stem that stores food and water; its nodes are called eyes.

Tuberous root: A thick, fibrous root that stores water and nutrients.

Wildflower: An herbaceous (nonwoody) annual, biennial, or perennial flowering plant capable of reproducing and becoming established without cultivation.

RECOMMENDED READING

Aiken, George D. *Pioneering with Wildflowers.* Brattleboro, Vt.: Alan C. Hood & Co., 1994.

Art, Henry W. *A Garden of Wildflowers: 101 Native Species and How to Grow Them.* Pownal, Vt.: Storey Communications, 1986.

Art, Henry W. *The Wildflower Gardener's Guide: California, Desert Southwest & Northern Mexico Edition.* Pownal, Vt.: Storey Communications, 1990.

Art, Henry W. *The Wildflower Gardener's Guide: Midwest, Great Plains & Canadian Prairies Edition.* Pownal, Vt.: Storey Communications, 1991.

Art, Henry W. *The Wildflower Gardener's Guide: Northeast, MidAtlantic, Great Lakes & Eastern Canada Edition.* Pownal, Vt: Storey Communications, 1987.

Art, Henry W. *The Wildflower Gardener's Guide: Pacific Northwest, Rocky Mountain & Western Canada Edition.* Pownal, Vt: Storey Communications, 1990.

Birdseye, Clarence, and Eleanor G. Birdseye. *Growing Woodland Plants.* New York: Dover Publications, 1972.

Bruce, Hal. *How to Grow Wildflowers and Wild Shrubs and Trees in Your Own Garden.* New York: Alfred A. Knopf, 1976.

Cobb, Boughton. *A Field Guide to Ferns and Their Related Families: Northeastern & Central North America.* (Peterson Field Guide Series). Boston: Houghton Mifflin Co., 1975.

Curtis, Will C. *Propagation of Wildflowers.* Revised by William E. Brumback. Framingham, Mass.: New England Wildflower Society, 1996.

Dana, Mrs. William Starr. *How to Know the Wildflowers.* Rev. ed. New York: Dover Publications, 1963.

Diekelmann, John, and Robert Schuster. *Natural Landscaping.* New York: McGraw-Hill Book Co., 1982.

Ferreniea, Viki. *Wildflowers in Your Garden.* New York: Random House, 1993.

Foster, F. Gordon. *Ferns to Know and Grow.* Portland, Ore.: Timber Press, 1993.

Hoshizaki, Barbara Joe. *Fern Growers Manual.* New York: Alfred A. Knopf, 1979.

Holmes, Roger, and Frances Tenenbaum, eds. *Taylor's Guide to Natural Gardening.* Boston: Houghton Mifflin Co., 1993.

Hull, Helen S., ed. *Gardening with Wildflowers.* (Plants & Gardens Series.) Brooklyn, N.Y.: Brooklyn Botanic Garden, 1962.

Imes, Rick. *Wildflowers.* Emmaus, Pa: Rodale Press, 1992.

Jones, Samuel B., Jr., and Leonard E. Foote. *Gardening with Native Wild Flowers.* Portland, Ore.: Timber Press, 1991.

Martin, Laura C. *The Wildflower Meadow Book.* 2nd ed. Chester, Conn.: The Globe Pequot Press, 1990.

Mickel, John T. *Ferns for American Gardens.* Indianapolis, Ind.: Macmillan Publishing USA, 1994.

Miles, Bebe. *Wildflower Perennials for Your Garden.* (American Garden Classics Series). Mechanicsburg, Pa.: Stackpole Books, 1996.

Newcomb, Lawrence. *Newcomb's Wildflower Guide.* Boston: Little Brown and Co., 1977.

Pauly, Wayne R. *How to Manage Small Prairie Fires.* Dane County Park Commission. Madison, Wisconsin. Reprinted 1988.

Peterson, Roger T., and Margaret McKenny. *A Field Guide to Wildflowers of Northeastern and North-Central North America.* (Peterson Field Guide Series.) Boston: Houghton Mifflin Co., 1975.

Phillips, Harry R. *Growing and Propagating Wild Flowers.* Chapel Hill, N.C.: University of North Carolina Press, 1985.

Phillips, Norma. *The Root Book: How to Plant Wildflowers.* Grand Rapids, Minn.: Little Bridge Publishing Co., 1984.

Rock, H.W. *Prairie Propagation Handbook.* 6th ed. Milwaukee, Wis.: Milwaukee County Department of Parks, 1981.

Sawyers, Claire, ed. *Gardening with Wildflowers and Native Plants.*(Plants & Gardens Series.) Brooklyn, N.Y.: Brooklyn Botanic Garden, 1989.

Smith, J. Robert., and Beatrice S. Smith. *The Prairie Garden.* Madison, Wis.: University of Wisconsin Press, 1980.

Snyder, Leon C. *Native Plants for Northern Gardens.* Chanhassen, Minn.: Andersen Horticultural Library, 1991.

Sperka, Marie. *Growing Wildflowers: A Gardener's Guide.* New York: Harper and Row Publishers, 1973.

Steffek, Edwin F. *The New Wild Flowers and How to Grow Them.* Enlarged and revised. Portland, Ore.: Timber Press, 1983.

Wasowski, Sally, and Andy Wasowski. *Gardening with Native Plants of the South.* Dallas, Tex.: Taylor Publishing Co., 1994.

Wilson, Jim. *Landscaping with Wildflowers.* Boston: Houghton Mifflin Co., 1992.

Wilson, William H. *Landscaping with Wildflowers and Native Plants.* San Francisco, Calif.: Ortho Books, 1984.

MAIL-ORDER SOURCES

This list includes mail-order sources for wildflower plants and seeds. Many of these companies are small businesses that must carefully watch costs. Most charge a small fee for their catalogs (generally one or two dollars). They often credit the fee toward your first order. Most companies also appreciate a self-addressed stamped business envelope enclosed with your inquiry.

Appalachian Wildflower Nursery
723 Honey Creek Road
Reedsville, PA 17084

Beersheba Wildflower Garden
P.O. Box 551
Beersheba Springs, TN 37305

Boothe Hill Wildflowers
23B Boothe Hill Road
Chapel Hill, NC 27514

Clyde Robin Seed Company
P.O. Box 2366
Castro Valley, CA 94546

Kurt Bluemel, Inc.
2740 Greene Lane
Baldwin, MD 21013

Eco-Gardens
P.O. Box 1227
Decatur, GA 30031

Forestfarm
990 Tetherow Road
Williams, OR 97544

Landscape Alternatives, Inc.
1705 Saint Albinos Street
Roseville, MN 55113

Midwest Wildflowers
Box 64
Rockton, IL 61072
Seed only

Missouri Wildflower Nursery
9814 Pleasant Hill Road
Jefferson City, MO 65109

Native Gardens
5737 Fisher Lane
Greenback, TN 37742

The Natural Garden
38W443 Highway 64
St. Charles, IL 60175

Niche Gardens
1111 Dawson Road
Chapel Hill, NC 27516

Plants of the Southwest
Route 6, Box 11A
Santa Fe, NM 87501

Prairie Moon Nursery
Route 3, Box 153
Winona, MN 55987

Prairie Nursery, Inc.
P.O. Box 306
Westfield, WI 53964

Prairie Restoration
P.O. Box 327
Princeton, MN 55371

Prairie Ridge Nursery/CRM
Ecosystems, Inc.
R.R. 2, 9738 Overland Road
Mt. Horeb, WI 53572

Putney Nursery, Inc.
Route 5
Putney, VT 05346

Sunlight Gardens
174 Golden Lane
Andersonville, TN 37705

Transplant Nursery
1586 Parkertown Road
Lavonia, GA 30553

The Vermont Wildflower Farm
P.O. Box 5
Route 7
Charlotte, VT 05445

We-Du Nurseries
Route 5, Box 724
Marion, NC 28752

Wildginger Woodlands
P.O. Box 1091
Webster, NY 14850

Wildseed Farms, Inc.
1101 Campo Rosa Road
P.O. Box 308
Eagle Lake, TX 77434

Woodlanders, Inc.
1128 Colleton Avenue
Aiken, SC 29801

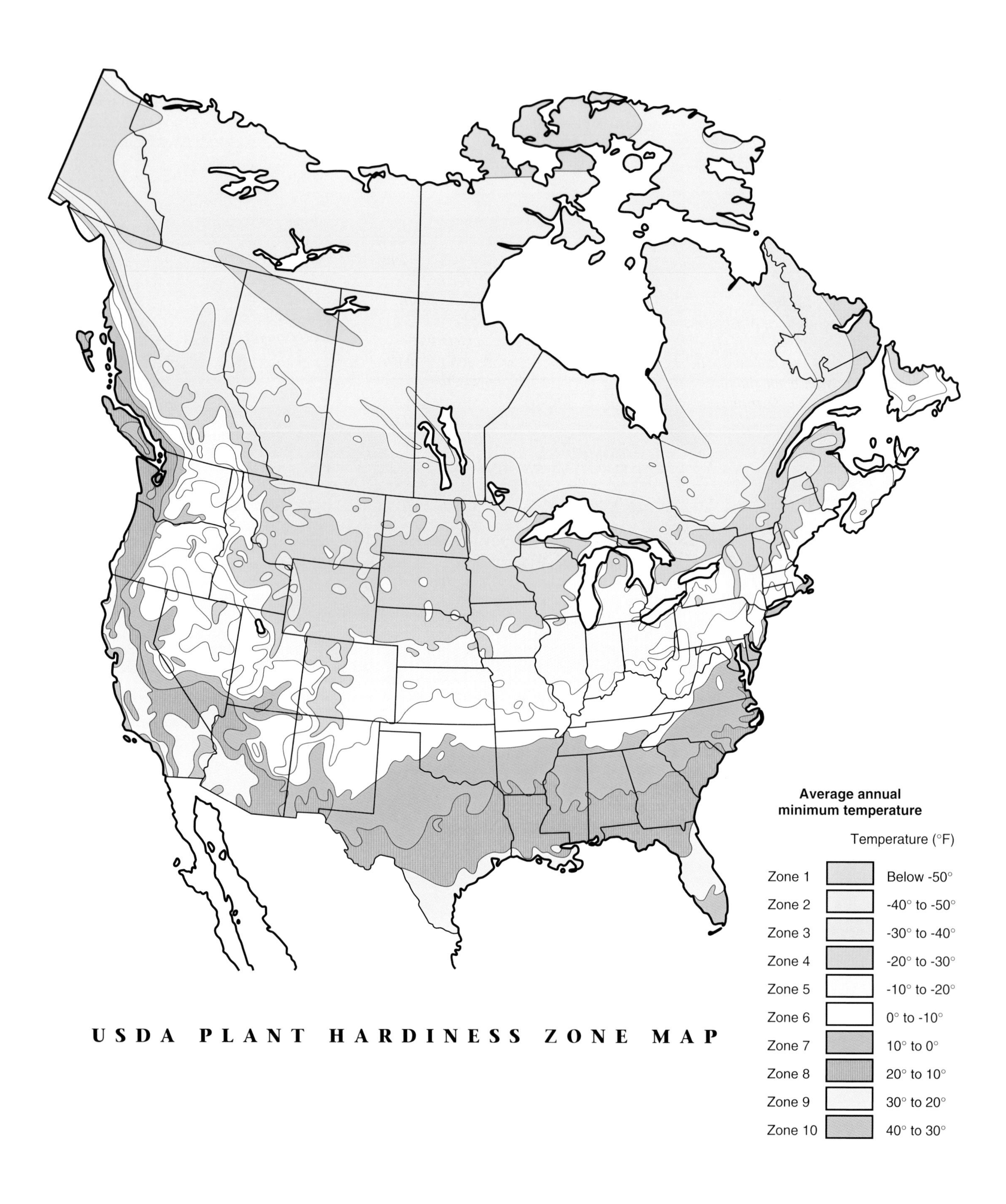

USDA PLANT HARDINESS ZONE MAP

COMMON AND BOTANICAL NAMES INDEX

INDEX

Photography Credits

Front Jacket Photograph: ©Adam Jones
Back Jacket Photography (counter clockwise from top left): ©Alan Detrick; PHOTO/NATS,INC.: ©Hal Horwitz; ©Alan and Linda Detrick; DEMBINSKY PHOTO ASSOCIATION: ©Joe Sroka; ©Alan Detrick; ©Adam Jones

©C. Colston Burrell: pp. 11 left, 11 right, 40, 41, 43, 44, 46, 47, 48, 49, 50, 51, 52, 53, 56, 62, 64, 68, 70, 74 left, 76, 78, 79, 82, 85, 86, 87, 88, 97, 101, 103, 104 right, 108, 109, 111, 112, 118, 119, 128, 129, 130, 132 left, 132 right, 137, 138, 139, 140, 142, 143, 147, 148, 150, 151, 153, 155, 156, 162, 165, 168, 172, 173 left

©Crandall & Crandall: p. 114

©R. Todd Davis: p. 19

©Richard Day/Daybreak Imagery: p. 106

©Alan and Linda Detrick: pp. 57, 93, 104 left

DEMBINSKY PHOTO ASSOCIATION: ©Willard Clay: p. 152; ©John Gerlach: pp. 105, 160; ©Ed Kanze: p. 136; ©Bill Lea: p. 83; ©Doug Locke: pp. 81, 98; ©Gary Mesazros: pp. 113 top, 113 bottom; ©John Mielcarek: p. 74 right; ©Skip Moody: p. 110; ©Jim Roetzel: p. 121; ©Dick Scott: p. 141; ©Joe Sroka: p. 92

ENVISION: ©B.W. Hoffmann: p. 18 left

©Derek Fell: pp. 14, 15 top, 61, 63, 67, 90, 94, 95, 161, 164

©John Glover: pp. 18 right, 89, 154

©Jessie M. Harris: pp. 42, 127, 135

©Bill Johnson: pp. 60, 133

©Adam Jones: pp. 8, 15 bottom, 17, 38, 177

©Dency Kane: pp. 13 right, 20, 29 top, 122

©Charles Mann: pp. 75, 99, 116, 123, 126

©Ed Monnelly: p. 100

©J. Paul Moore: pp. 54, 59, 71, 125

Niche Gardens: ©Kim Hawks: p. 96

©Clive Nichols: p. 73

©Jerry Pavia: pp. 4, 13 left, 16, 84, 120, 146

©Joanne Pavia: pp. 72, 115

PHOTO/NATS,INC.: ©Gay Bumgarner: p. 166; ©Deborah M. Crowell: p. 91; ©Wally Eberhart: p. 65; ©Hal Horwitz: pp. 77, 117, 131; ©Sydney Karp: p. 134; ©Bill Larkin: p. 145; ©John A. Lynch: p. 69; ©A.Peter Margosian: p. 58; ©Herbert B. Parsons: p. 32 top; ©David M. Stone: pp. 144, 158

©Kenneth E. Smiegowski: pp. 33 top, 173 right

©Richard Shiell: pp. 10 left, 12

TOM STACK ASSOCIATES: ©Rod Planck: p. 33 bottom

©John J. Smith: pp. 10 right, 163

©VISUALS UNLIMITED: ©David S. Addison: p. 149; ©Pat Armstrong: p. 31 left; ©Callahan,S.: p. 102; ©Gary W. Carter: pp. 107, 124; ©John Gerlach: p. 45; ©Bill Johnson: p. 159; ©H.A. Miller: p. 66; ©John Sohlden: p. 55; ©Richard Thom: pp. 80, 157; ©William J. Weber: p. 32 bottom

©judywhite/New Leaf Images: pp. 30, 31 right

Illustration Credits

C. Colston Burrell: pp. 171, 175, 179

Frank Fretz: pp. 22 top, 22 bottom, 23, 24, 25, 26 top, 26 bottom left, 26 bottom right, 27, 28, 29 bottom, 34, 35, 36, 37 left, 37 right, 180, 181, 182